DESIGN TEXTBOOKS
IN CIVIL ENGINEERING: VOLUME VII

ARCHES AND SHORT SPAN BRIDGES

DESIGN TEXTBOOKS IN CIVIL ENGINEERING: VOLUME VII

ARCHES AND SHORT SPAN BRIDGES

SERGE LELIAVSKY

Ph.D., M.I.C.E., F.Am.Soc.C.E.

formerly Director, Designing Service, Reservoirs and Nile Barrages Department, Ministry of Public Works, Cairo, Chief of Designing Office, Irrigation Projects Department, Ministry of Public Works, Cairo, Professor of Irrigation Design, Royal School of Engineering, Guiza, Cairo, Head of the Bridges Department, Egyptian State Railways, Cairo, Head of the Hydraulic Section of Lock and Dam Construction Works on the Dneiper Falls, Russia

1982
LONDON   NEW YORK
CHAPMAN AND HALL

First published 1982 by
Oxford & IBH Publishing Co
66 Janpath, New Delhi, India
and
Chapman and Hall Ltd
11 New Fetter Lane, London EC4P 4EE

Published in the USA by
Chapman and Hall
in association with Methuen, Inc
733 Third Avenue, NY 10017

© *1982 Chapman and Hall Ltd*

ISBN 0 412 22560 3

All rights reserved. No part of this book
may be reprinted, or reproduced or utilized
in any form or by any electronic, mechanical
or other means, now known or hereafter invented,
including photocopying and recording, or in any
information storage or retrieval system, without
permission in writing from the Publisher

Printed in India by Oxonian Press Pvt. Ltd.,
Faridabad

AUTHOR'S PREFACE

Like the earlier volumes in this series (*Design Textbooks in Civil Engineering*), the present book has been abstracted from and based upon the author's previous work, *Irrigation and Hydraulic Design*.

The general objective in producing this volume was to place in the hands of the student at a relatively low price the knowledge required for the design of a particular type of hydraulic work, whereas the earlier publication, of much wider scope, was of necessity more expensive.

The reason for selecting arches and bridges of short span, to be dealt with in a single volume, was that these types constitute, in practice, the ingredients of a very large variety of major civil structures, such as hydraulic and hydro-electric head-works, dams, barrages and weirs, bridges and valley-crossings, etc. The civil engineer, engaged in all these works, will constantly meet with these two types in his daily routine work. A design book dealing with them has therefore a good chance of remaining permanently to hand on the designer's desk.

In conclusion, the author again wishes to record his appreciation of the excellent work done by the publishers in producing these books.

S. L.
1963

CONTENTS

1. Arches   page 1

Section I. STRESSES   1

1. Introductory Notes   1
2. Deflection of Curved Beams   5
3. Further Analysis of the Three Fundamental Elastic Equations   9
4. Solution of the Elastic Equations in Practice   15
5. Effect of Temperature   27

Section II. DESIGN PROCEDURE   33

1. The Problem   33
2. Melan's Method   38
3. Kögler's Method   42
4. Cochrane's Method   45

2. Timber Bridges   48

Section I. GENERAL REMARKS   48

Section II. DESIGN OF TIMBER BRIDGES   48

1. Introductory   48
2. Details of Design of Timber-work, Carrying Calculated Stress   48
3. Application of Main Principles to the Design of Timber Structures   51
4. Types of Standard Connections   54
5. Trestles and Supports   57

Section III. EXAMPLE OF DESIGN OF A TIMBER SPAN   59

1. Flooring, Footpath, etc.   59
2. Main Girders   63
3. Spacing of Keys   66

Section IV. STRENGTHENING AN EXISTING TIMBER BRIDGE   71

Section V. NEW TENDENCIES IN STRUCTURAL TIMBER DESIGN   73

1. Introductory   73
2. "Connectors"   73
3. Golden Gate International Exposition in San Francisco   74
4. Plywood I-beams   77

3. Rolled-joist Bridges — 84

Section I. TWO TYPES — 84

Section II. TYPE A—ROLLED-JOIST BRIDGES WITH TIMBER FLOORING — 84

1. Pros and Contras — 84
2. Calculations — 84
3. Various Designs — 84
4. Bearings and Base-plates — 90

Section III. TYPE B—ROLLED-JOIST BRIDGES WITH REINFORCED-CONCRETE SLAB — 91

1. Pros and Contras — 91
2. Calculations — 91
3. Three Subclasses — 91

Section IV. CALCULATION OF SLAB — 91

1. General Order of Procedure — 91
2. "French" Formula — 92
3. "American" Formula — 93
4. "German" Method — 93
5. Comparison and Example of Application — 93
6. Required Thickness of Slab — 94
7. Checking the Stress — 99
8. The Shear — 100
9. Application of Different Formulae — 101
10. Conclusions — 101

Section V. STRUCTURAL DETAILS OF SLAB — 102

1. General Remark — 102
2. Ends of Bars — 102
3. Junction of Bars — 102
4. Bends — 104
5. Under-surface of Slab — 104
6. Bearings — 104
7. Bracings — 105
8. Special Types — 106

Section VI. ROLLED-JOIST-BRIDGES, IN WHICH STEEL AND CONCRETE PARTICIPATE IN CARRYING BENDING MOMENT — 107

1. Introductory — 107
2. Earlier Designs *versus* Modern Tendencies — 107
3. Shear "Connectors" — 109
4. Patented Continuous "Connectors" — 110

4. Reinforced-concrete Bridges — 111

Section I. GENERAL PRINCIPLES — 111

CONTENTS

    1. T-beam Bridges Only   111
    2. Reader Supposed to be Acquainted with General Principles of R.C. Design   111
    3. Limits of Spans for Various Types   111
    4. Steel Percentages   112

Section II. EXAMPLE OF R.C. BRIDGE. DESIGN OF SLAB   112

    1. General Notes   112
    2. Spacing Cross-girders   113
    3. Calculation of Moments for Design of Slab   114
    4. Formulae for Slabs Supported on Four Sides   115
    5. Final Values for Bending Moments and Corresponding Design of Slab   115
    6. Adopting Low Stress Limits ($f_0 = 35$ kg/cm² and $f_s = 900$ kg/cm²)   116
    7. Argument in Favour of $n = 12$ 117
    8. Footpath   117

Section III. DESIGN OF CROSS-GIRDERS   118

    1. Introductory   118
    2. Dead-load Moments   118
    3. Live-load Moments   119
    4. Design of Section   120
    5. Checking the Stresses   123
    6. Stirrups   126

Section IV. DESIGN OF MAIN GIRDERS   128

    1. Order of Procedure in General   128
    2. Analytical Formulae   129
    3. Graphical Calculation—Method Recommended   129
    4. Approximate Analytical Solution, Convenient for Preliminary Analysis   131
    5. Calculation of Bending Moments   131
    6. Preliminary Design, by Means of Approximate Formula   132
    7. Verification by Graphical Method   133
    8. Question of Scales   135
    9. Bent Bars and Spacing of Stirrups   136
    10. Checking Lengths of Bars, with Reference to Bending Moment Diagram   138

5. Pre-stressed Bridges   140

Section I. GENERAL THEORY OF PRE-STRESS   140

    1. The Interdependence of the Advantages of Pre-stress and Span   140
    2. Principle of Pre-stress   140
    3. Basic Idea of Pre-stress Not New   142
    4. Continuous and Discontinuous Structures   143
    5. Non-elastic Behaviour of Materials   144
    6. Bonded and Non-bonded Types of Pre-stress   145
    7. Comparison with a Retaining Wall   146

Section II. CALCULATION OF A PRE-STRESSED BEAM   149

1. The Assumption of Non-cracked Concrete ... 149
2. Effect of Cracks ... 152
3. Effect of 'Creep' ... 155

Section III. SKEW SPANS ... 157

1. General Considerations ... 157
2. Examples ... 157

Section IV. POPULAR TYPES FOR PRE-STRESSED BRIDGES OF MODERATE SPAN ... 159

1. Solid Slab ... 159
2. Hollow Slab ... 160
3. I-beams ... 160
4. Channel-slab and Hollow Stringer ... 160
5. Combination of Pre-cast Elements with Cast-*in-situ* Parts ... 161

Section V. GIFFORD TYPE ... 161

6. Specifications for Steel Bridges ... 166

Section I. EARLIER HISTORY ... 166

1. Foreword ... 166
2. Intensities of Working Stresses in General ... 166
3. Permissible Intensity for Tensile Stress ... 167
4. Wöhler's Tests ... 171
5. Launhardt-Weyrauch Formula ... 172

Section II. LATER PERIOD ... 173

1. Methods of Dealing with Impact Effect ... 173
2. The Fereday-Palmer Instrument ... 177
3. Electrical Measuring Instruments ... 179
4. Impact for Highway Bridges ... 182
5. Impact Effect Dealt with as a Function of the Speed of Vehicles ... 185

Section III. SPECIAL ALLOY STEELS ... 186

1. Different Kinds of Steels Used in Parallel in the Same Bridge ... 186
2. Examples of Bridges in which Special Alloy Steel was Used ... 187
3. Permissible Stress Intensities for Special Alloy Steels ... 188
4. Various Considerations Affecting the Use of Special Alloy Steels ... 189
5. Special Alloy Steel in the Bridges and Chelsea of Howrah. Question of Rivets ... 189
6. Limitations for the Use of Alloy Steels ... 190
7. Alloy Steels and Corrosion ... 191

Section IV. PHILOSOPHY OF SAFETY FACTOR ... 192

1. Generalized Concept of Safety Factor ... 192
2. Three Types of Error ... 193
3. Solutions Based on Mathematical Theory of Probability ... 194

CONTENTS xi

 Section V. LIMIT DESIGN AND PLASTIC THEORIES   198

  1. Simple Examples Explaining Principle of Limit Design and Plastic Theories   198
  2. Deeper Inquiry into the Beam Failure   200
  3. W. Kuntze's Empirical Method and Other Theories   201
  4. Vertical Bar Flanked by Two Sloping Laterals   203
  5. Rigid Horizontal Beam Suspended on Three Wires   205

 Section VI. LIMIT STRESS FOR COMPRESSION   207

  1. Danger of Buckling. Schwarz-Rankine Formula   207
  2. Net and Gross Areas   208
  3. Straight-line Formula   208
  4. Euler's Formula   208
  5. Failure of Old Quebec Bridge   209
  6. Buckling of Upper Flanges in Plate Girders   210

 Section VII. LIMITS FOR SHEAR AND BEARING STRESSES   211

  1. Shear   211
  2. Bearing Stress   212

7. Plate-girder Bridges   213

 Section I. GENERAL ORDER OF PROCEDURE   213

 Section II. PLATE-GIRDERS IN GENERAL   215

  1. Two Types: Continental and Anglo-American   215
  2. Continental Method   216
  3. Anglo-American Method   221
  4. Comparison of the Two Types   223

 Section III. RAIL BEARERS AND CROSS-GIRDERS   225

  1. Design of Rail Bearers   225
  2. Design of Cross-girders   231

 Section IV. CALCULATION OF THE MAIN GIRDERS   237

  1. Standard Type   237
  2. Continuous Main Girders   238
  3. Precise Stress Calculation for Continuous Plate-girder Bridges   242
  4. Structural Design of the Main Girders   245

 Section V. SECONDARY DETAILS: BEARINGS, EXTREME SLEEPERS, BRACINGS, ETC.   248

  1. Bearings   248
  2. Details of Extreme Sleepers   252
  3. Wind Bracings   253

 Section VI. OTHER TYPES OF PLATE-GIRDER BRIDGES   254

  1. Railway Plate-girder Bridges of Other Types   254
  2. Plate-girder Road Bridges   256

xii   CONTENTS

    Section VII. RIVETS AND WELDS 260

        1. General Rules for Design of Riveted Connections 260
        2. Welded Connections 262

8. Movable Bridges 264

    Section I. LIFT BRIDGES 264

        1. General Principle of a Lift Bridge 264
        2. Location of Centre of Rotation 265
        3. Main Shaft 265
        4. Bascule Type in Germany and America 267
        5. The Strauss Bridge 267
        6. The Scherzer Bridge 269
        7. Egyptian Irrigation Type 269
        8. Belidor or Sinusoid Type 270

    Section II. SWING BRIDGES 273

        1. Two Types: Pivot-bearing and Rim-bearing 273
        2. Wedging Gear 275
        3. Pros and Contras of the Four Systems 276
        4. Unusual Solutions 284
        5. Fenders 285

9. Aqueducts 286

    Section I. INTRODUCTORY 286

    Section II. HYDRAULIC DESIGN 286

    Section III. STRUCTURAL DESIGN 288

        1. General Notes 288
        2. Calculation of the Bending Moment 288
        3. Calculation of the Thickness of the Metal and Design of Joints ... 290
        4. Anchorage 294

SUBJECT INDEX 295

AUTHOR INDEX 297

Chapter One

ARCHES

SECTION I. STRESSES

1. Introductory Notes

Just as in many other engineering types, the unit stresses in an arch are functions of bending moments, and axial and shearing forces. It follows, therefore, that in order to calculate the stresses in an arch, it is first necessary to determine all the forces and moments applied to it.

Viewed from this standpoint, the arch will be found to possess the same mechanical characteristics as a curved beam, with fixed ends. The forces and moments applied to such a beam are then as follows (see Fig. 1):

(1) External forces P_1, P_2, P_3, etc., including live loads and structural weight.
(2) Vertical reaction V at the left-hand skewback.
(3) ,,   ,, V_1 ,, ,, right-hand ,,
(4) Horizontal reaction or thrust H at the left skewback.
(5) Horizontal thrust H_1 at the right skewback.
(6) Moment M at the left skewback.
(7)   ,, M_1 ,, ,, right   ,,

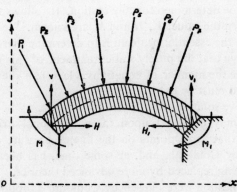

Fig. 1.

In problems concerning the stability of arches, the external forces P_1, P_2, P_3, etc., are supposed to be known. For the determination of the reactions, the theory of structural statics supplies three equations, namely:

$$\Sigma F \cos \zeta = 0 \qquad \text{(I)}$$
$$\Sigma F \sin \zeta = 0 \qquad \text{(II)}$$
$$\Sigma M_F = 0 \qquad \text{(III)}$$

where $\Sigma F \cos \zeta$ and $\Sigma F \sin \zeta$ represent respectively the two components, parallel to the axes of co-ordinates, of the resultant of all the forces; and ΣM_F is the sum of all moments.

The particular point of the problem of the calculation of stresses in an arch lies in the insufficiency of the number of static equations for the determination of the reactions; in fact, we have six unknowns, V, V_1, H, H_1, M and M_1, as against three equations of equilibrium only (I, II and III).

The better to explain this point, suppose that the values V, H and M are known. Then $R = \sqrt{V^2 + H^2}$ is the resultant reaction and $\dfrac{V}{H} = \tan \delta$ is the tangent of the angle between this resultant and the vertical (see Fig. 2).

Let ab be a vertical line drawn through the centre C of the skewback. Then $\varDelta = \dfrac{M}{H}$ represents the distance from C to the point where this line intersects the reaction R.

Thus, the three values V, H and M are sufficient to determine the magnitude and the direction of the reaction R, and when this reaction is found, the resultant of the forces acting on any cross-section of the arch can be easily obtained by means of a stress diagram, as explained below.

The resultant for a section taken in between the forces P_1 and P_2 is the sum S_1 of the reaction R and the force P_1; it follows that, by drawing through e_1 a parallel to S_1 we can get the element of the pressure polygon between the points e_1 and e_2. In the same way, the resultant of the forces S_1 and P_2, which, in the stress diagram, is marked by the letter S_2, is the resultant for any cross-section taken between P_2 and P_3. The element of the line of pressure between e_2 and e_3 is, therefore, parallel to S_2. Following the same method, the line of pressure may be computed for the whole length of the arch, and thus, the position and value of the corresponding resultant may be determined for any section. It follows that all the stresses in an arch can be calculated when the values V, H and M are known. We may, therefore, conclude that in calculating an arch the essential problem is to determine these three values.

We have already shown that the purely static treatment of this question does not supply the required result because the number of equations is less than the number of unknown values; hence, another method must be used.

In fact, three alternative solutions may be followed:

(a) The calculation may be based upon three additional arbitrary assumptions; for instance, we may assume that three points on the line of pressure are known. This method has first been suggested by Coulomb, and, at some time, has been much used; at present, however, it is gradually being replaced by more advanced theories.

The three points in question were chosen by Coulomb as follows: in the upper limit of the middle third at the crown, and in the lower limits of the middle thirds at the springings (see Fig. 3).

Fixing *a priori* one point on the line of pressure is equivalent to one arbitrary equation. Three points are therefore necessary for compensating the three missing equations. It may easily be shown that should these three points on the line of pressure be known, or supposed to be known, we could then draw the whole line and calculate the three unknown values V,

ARCHES

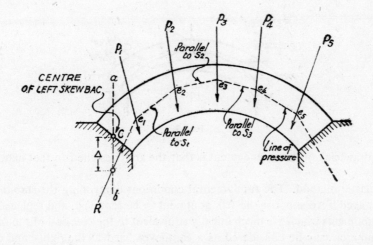

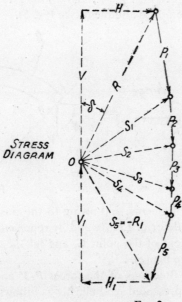

Fig. 2.

H and M; which would then allow the stresses to be found.

There is no proof whatever to show that this method is correct, but it is often adopted as an empirically established criterion for short-span arches and current conditions of loading.

(b) The second method consists in designing the arch in such a way as to force artificially the line of pressure to pass through three chosen points. This leads to the, so-called, three-hinged arch (see Fig. 4). The calculations are made in the same way as before, but they are then developed from a correct physical basis, instead of an illusory assumption. On the other hand, this type of design is expensive and is used only in the case of heavy arches and large spans.

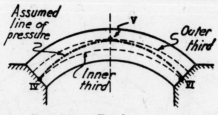

FIG. 3.

Another drawback of the arrangement is that the arch designed in that manner may lack rigidity.

(c) *The elastic method.* The fundamental conditions controlling the stability of an arch will not be changed if we suppose the left abutment to be removed, and replaced by a system of forces and moments which are mechanically equivalent to the reaction of the removed abutment. Thus, an arch may be considered as a cantilever fixed at the right-hand end, and subjected to the action of (a) the external forces P_1, P_2, etc., and (b) the forces V and H and the moment M replacing the reaction of the removed abutment (see Fig. 5).

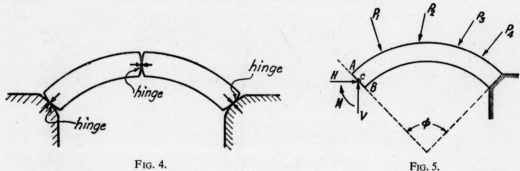

FIG. 4. FIG. 5.

Under the effect of these forces and moments, and owing to the postulated elasticity of the structural material, the cantilever will deform. Let Δx and Δy represent the horizontal and vertical components of the elastic movement of the point C, and let $\Delta \phi$ represent the elastic angular deflection of the line AB (see same figure).

It is evident that Δx, Δy and $\Delta \phi$ are functions of the forces P, V and H and of the moment M. This general principle may be represented symbolically as follows:

$$\Delta x = f_1(P, V, H, M)$$
$$\Delta y = f_2(P, V, H, M)$$
$$\Delta \phi = f_3(P, V, H, M)$$

But, as in Nature the abutments are solid, and are, therefore, assumed in the usual calculations to be practically rigid,[1] the unknown forces V and H and the unknown moment M must be such that

$$\Delta x = 0$$
$$\Delta y = 0$$
$$\Delta \phi = 0$$

---

[1] The theory of elastic abutments and piers forms a separate branch of the general theory of arches, which is not discussed in these pages.

ARCHES

Consequently we obtain three equations:

$$f_1(P, V, H, M) = 0 \qquad \text{(IV)}$$
$$f_2(P, V, H, M) = 0 \qquad \text{(V)}$$
$$f_3(P, V, H, M) = 0 \qquad \text{(VI)}$$

These are the three additional, so-called "elastic", equations, which are required to solve the problem.

In combination with the three "static" equations (I, II and III) quoted earlier, they supply the complete solution of the problem.

The elastic theory embodied in the three equations, IV, V and VI, though it may require more computation-work than any other method, has proved in practice to be the only fully reliable method applicable to a wide variety of conditions of loading and design.

Its practical weight was demonstrated, for the first time, in the well-known experiments carried out, with full-size arches, as well as with small-scale models, by the Austrian Union of Engineers and Architects, in 1889–92, and has since been confirmed by thousands of instances, in which the method was successfully applied. This solution is now universally adopted for the design of all important masonry and concrete arches. The conventional methods, based on arbitrary assumptions, are, however, sometimes used for very short-span bridges, as empirical solutions.

2. Deflection of Curved Beams

Sketch I in Fig. 6 represents the stress diagram for a section of a straight beam, subject to bending. The stresses in the extreme fibres of the beam are $\eta_1 = \eta_2 = \dfrac{M_T B}{2I}$, where M_T is the total bending moment. According to the usually adopted principles of the theory of stresses, the elastic deformation of the fibres of the element $abcd$ is graphically represented by the straight line a_1b_1 (Sketch 1), the stretch being equal to the stress, η, multiplied by the length, ds, of the element, and divided by the modulus of elasticity E (Young's Modulus), so that

$$e = \frac{M_T B ds}{2IE}.$$

Further we have: $\tan \theta = \dfrac{\overline{aa_1}}{\frac{1}{2}ab} = \dfrac{2e}{B}$. As for very small angles $\tan \theta$ may be assumed to be equal to θ, we may also write:

$$\theta = \frac{2e}{B} = \frac{M_T ds}{IE}.$$

Turning, now, to the case of a curved beam, which is represented in Sketch II in the same figure, we shall find that each elementary fibre has a different length. Assuming ds to represent the infinitely small length of such a fibre, measured parallel to the axis of the beam between two normal planes ab and cd, we shall have

$$ac = \frac{r + 0.5B}{r} ds$$

$$bd = \frac{r - 0.5B}{r} ds.$$

ARCHES AND BRIDGES

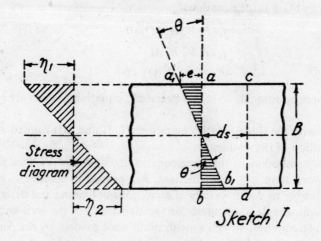

Sketch I

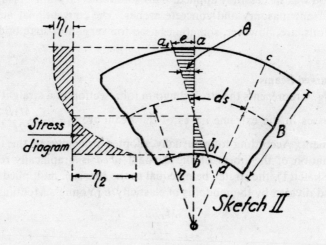

Sketch II

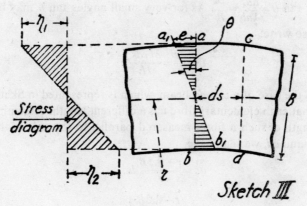

Sketch III

Fig. 6.

ARCHES

It is commonly assumed that in curved beams, as in straight ones, the planes ab and cd remain very nearly straight, after the bending has taken place; but, owing to the differences in the lengths of the elementary fibres, the stress diagram in the case of a curved beam is no longer a straight line, but a curve; namely, for rectangular beams, a logarithmic hyperbola, as represented in the sketch. Accordingly, for curved beams the stresses η_1 and η_2 are not equal, the larger stress corresponding to the smaller elementary length of fibre, and vice versa. For instance, in the case of a hook (see Fig. 7) the stress in the point b is materially greater than in the point a, a fact which is well known to machine designers.

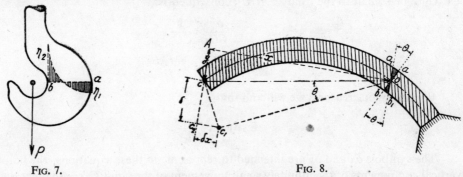

FIG. 7.   FIG. 8.

The difference $\eta_2 - \eta_1$ depends on the difference in the lengths of fibres, i.e. on

$$ac - bd = \left(\frac{r + 0.5B}{r} - \frac{r - 0.5B}{r} \right) ds = \frac{B}{r} ds.$$

Nevertheless, both calculations and experiments concur in showing that for design purposes ac can be assumed equal to bd, if and when $\dfrac{B}{r} < \dfrac{1}{4}$.

For arches this criterion is nearly always satisfied.

This explains why it is that, in practice, the distribution of stresses in an arch is currently assumed to be linear; in fact, this assumption means that the case is assimilated to that of a straight beam (see Sketch III in same figure), and θ is therefore taken to be equal to $\dfrac{M_T ds}{IE}$.

Let us now suppose that ab is an element of an arch (see Fig. 8). Also, let it be assumed that the right-hand end of this arch is rigidly fixed into the abutment, while its left-hand end is left free to move. Under such assumptions it will follow that the elastic distortion of the element considered, will cause all the part of the beam placed to the left of the said element to change its position: that is to say, this part will swing around the centre of the element, as if the latter were a hinge; the elastic angle of rotation being $\dfrac{M_T ds}{IE}$.

Thus, the inclination of the extreme element at the left-hand extremity of the arch will be the sum of the elastic distortions of all the intermediate elements, i.e.

$$\Delta\phi = \int_{s=0}^{s=S} \frac{M_T ds}{IE}$$

where S is the total developed length of the axis of the arch.

Further, let the co-ordinates of the centre, O, of the section AB, be x and y, as shown in the same figure. Then, whilst one element only is distorted, the path travelled by the point C, from its original position to C_1, will be equal to

$$CO \times \theta = \frac{CO \times M_T ds}{IE}.$$

For very small angles θ, it may be assumed that CC_1 is straight and perpendicular to CO. Accordingly, the angle $\angle C_1 C C_2$ may be taken to be equal to the angle $\angle AOC$, and the triangle $CC_1 C_2$ to be similar to the triangle AOC_1; consequently:

$$\frac{\delta x}{CC_1} = \frac{y}{CO}.$$

$$\delta x = y \frac{CC_1}{CO} = y \frac{M_T ds}{IE}.$$

For analogous reasons, we will find that

$$\delta y = x \frac{M_T ds}{IE}.$$

The symbols δx and δy are intended to represent, in these equations, the horizontal and vertical components of the infinitely small movement of the point C, consequent on the elastic deformation of an infinitely narrow element ds of the beam.

The total movement of the point C is the combined result of the deformations of all such elements.

Hence, in order to obtain the components of the *total* displacement, we must integrate δx and δy between the limits $s=0$ and $s=S$. The horizontal component of the total displacement of the point C is therefore equal to

$$\int_{s=0}^{s=S} \delta x = \int_{s=0}^{s=S} \frac{M_T y}{IE} ds$$

while the vertical component of the same displacement is

$$\int_{s=0}^{s=S} \delta y = \int_{s=0}^{s=S} \frac{M_T x}{IE} ds.$$

Thus, the three fundamental equations of the elastic theory of arches are now found to be

$$0 = \Delta x = \int_0^S \frac{M_T y}{IE} ds \qquad (IV_1)$$

$$0 = \Delta y = \int_0^S \frac{M_T x}{IE} ds \qquad (V_1)$$

$$0 = \Delta \phi = \int_0^S \frac{M_T}{IE} ds. \qquad (VI_1)$$

ARCHES

It should be realized that, so far, we have not taken into account the elastic deformation of the beam due to normal forces; we have considered only the deflections due to the bending moment, neglecting the elastic deformation due to the compressive stress set up by the horizontal thrust and vertical reaction.

Strictly speaking, this additional deformation should also be taken into account, especially for very flat arches; but, as in this case the formulae become still more involved, the deflections due to the axial stresses are often neglected in practice. This results in overestimating the value of the horizontal thrust. According to Howe, the error is about 10 per cent, for arches having a ratio of rise to span = 1/5, and is increased to about 30 per cent when this ratio falls to 1/10. Hence, the simple formulae given above should not be applied in the latter case. However, since such flat arches are seldom used in irrigation and hydraulic design, the reader is referred to special courses for formulae covering the effect of H.

Note that the coefficient E, appearing in the deflection formula, is considered constant in the case of masonry or plain concrete arches. In the case of reinforced concrete arches the value I is calculated as the sum of the moment of inertia of the section of concrete plus twelve to fifteen times the moment of inertia of the section of steel. This automatically covers the effect of the difference in the elastic properties of steel as against concrete, and, consequently, for the calculation of reinforced concrete arches also, E may be taken to be constant.

In so far as mechanical loads are concerned, the value of E, provided it is constant, does not affect the result of the calculation. Thermal stresses will be considered later on.

3. Further Analysis of the Three Fundamental Elastic Equations

Let us assume that the positive direction of moments is as represented by the arrow in Fig. 9, i.e., clockwise.

The total moment, M_T, for any section, ab, can then be calculated as the algebraic sum of the following:

(a) $m =$ *Moment of external forces*. In the case shown in the drawing

$$m = -(P_1K_1 + P_2K_2 + P_3K_3).$$

The sign minus is included in this formula because the direction of the moments of the external forces, in this case, is inverse to that shown by the arrow. Note that only the forces placed to the left of the section ab are taken into account in calculating the local value of m.

In the case when distributed loads are applied to the arch, the moment m can either be

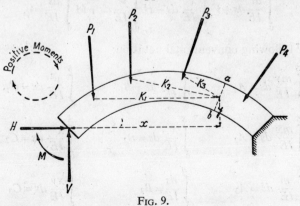

Fig. 9.

calculated analytically, or the distributed loading may be replaced by a series of equivalent concentrated loads, and the moment calculated numerically or graphically, as may appear more convenient to the designer. Note also that the moment m is the same as in a beam fixed at one end and free at the other.

(b) $V \times x =$ *Moment of the reaction V.* This moment will be included in the equation with the sign plus, in accordance with our assumption regarding the positive direction of moments.

(c) $-H \times y =$ *Moment of the reaction H.* In this case the direction of the moment is opposite to that shown by the arrow and, consequently, the minus sign is used.

(d) $M =$ *The reactive moment* (or the fixation moment) at the left skewback.

Thus, the total bending moment in any section ab of the arch is equal to

$$M_T = m + Vx - Hy + M.$$

Hence, equation (IV$_1$) may now be re-written as follows:

$$\int_0^S \frac{M_T y}{IE} ds = \int_0^S \frac{(m + Vx - Hy + M) y}{IE} ds$$

$$= \int_0^S \frac{my}{IE} ds + V \int_0^S \frac{xy}{IE} ds - H \int_0^S \frac{y^2}{IE} ds + M \int_0^S \frac{y}{IE} ds = 0. \tag{IV$_2$}$$

In the same manner equation (V$_1$) becomes:

$$\int_0^S \frac{M_T x}{IE} ds = \int_0^S \frac{(m + Vx - Hy + M) x}{IE} ds$$

$$= \int_0^S \frac{mx}{IE} ds + V \int_0^S \frac{x^2}{IE} ds - H \int_0^S \frac{yx}{IE} ds + M \int_0^S \frac{x}{IE} ds = 0. \tag{V$_2$}$$

The sixth equation is converted into

$$\int_0^S \frac{M_T}{IE} ds = \int_0^S \frac{m + Vx - Hy + M}{IE} ds$$

$$= \int_0^S \frac{m}{IE} ds + V \int_0^S \frac{x}{IE} ds - H \int_0^S \frac{y}{IE} ds + M \int_0^S \frac{ds}{IE} = 0. \tag{VI$_2$}$$

Let us adopt the following conventional notation:

$$\int_0^S \frac{my}{IE} ds = A_1 \qquad \int_0^S \frac{y}{IE} ds = B_1 \qquad \int_0^S \frac{y^2}{IE} ds = C_1$$

$$\int_0^S \frac{mx}{IE} ds = A_2 \qquad \int_0^S \frac{x}{IE} ds = B_2 \qquad \int_0^S \frac{x^2}{IE} ds = C_2$$

$$\int_0^S \frac{m}{IE} ds = A_3 \qquad \int_0^S \frac{ds}{IE} = B_3 \qquad \int_0^S \frac{xy}{IE} ds = C_3$$

ARCHES

The three "elastic" equations will then be as follows:

$$A_1 + VC_3 - HC_1 + MB_1 = 0 \qquad (IV_3)$$

$$A_2 + VC_2 - HC_3 + MB_2 = 0 \qquad (V_3)$$

$$A_3 + VB_2 - HB_1 + MB_3 = 0 \qquad (VI_3)$$

These three equations supply the full solution to the problem.

In point of fact, we have now three linear equations with three unknowns, V, H and M. All the coefficients in these equations can be calculated; and namely, the integrals B_1, B_2, B_3, C_1, C_2 and C_3 depend on the dimensions of the arch only; hence, they will not change, whatever the loading. The other three integrals, namely, A_1, A_2 and A_3, depend both on the shape and material of the arch, and also on the loads; consequently, they are recalculated every time the loading changes.

From the purely theoretical standpoint the calculation of these nine integrals should not lead to any particular difficulty. It can be performed using any of the numerous formulae and methods for numerical integration.

There is, also, no difficulty in solving three linear equations with three unknowns. Such calculations are, nevertheless, extremely laborious and require considerable time to carry out, and, since during the last fifty years the elastic method has gradually been adopted by all designers, numerous methods have been suggested with the object of simplifying the arithmetical calculation of the integrals, and the solution of the three main equations.

One of the most useful simplifications consists in replacing the three fundamental elastic equations by three others, every one of which contains *one* unknown only. For a symmetrical arch the latter three equations can be developed as explained below.

Let us replace M (the unknown moment at the skewback) by a new variable, namely:

$$\mathcal{M} = M + aV - bH$$

where a and b are two constants determined in such a manner that

$$a = \frac{\int_0^S \frac{x\,ds}{IE}}{\int_0^S \frac{ds}{IE}} \quad \text{and} \quad b = \frac{\int_0^S \frac{y\,ds}{IE}}{\int_0^S \frac{ds}{IE}}. \qquad (A)$$

Thus $$M = \mathcal{M} - aV + bH.$$

Further, let us change the origin of co-ordinates and shift it to the point $x = a$ and $y = b$ (see Fig. 10).

The new co-ordinates will then be determined as follows:

$$z = x - a \quad \text{and} \quad u = y - b.$$

Note that

$$\int_0^S \frac{z}{IE} ds = \int_0^S \frac{x}{IE} ds - a \int \frac{ds}{IE} = \int_0^S \frac{x\,ds}{IE} - \frac{\int_0^S \frac{x\,ds}{IE}}{\int_0^S \frac{ds}{IE}} \int_0^S \frac{ds}{IE} = 0.$$

In the same manner it may be shown that

$$\int_0^S \frac{u}{IE}\,ds = 0.$$

For a symmetrical arch a is equal to half the span, $\frac{l}{2}$, and

$$\int_0^{\frac{S}{2}} \frac{uz}{IE}\,ds = -\int_{\frac{S}{2}}^S \frac{uz}{IE}\,ds.$$

It follows that $\int_0^S \frac{uz}{IE}\,ds = 0.$

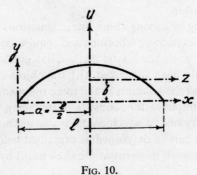

Fig. 10.

Including $\mathcal{M}$, z and u in equation (VI$_2$), we get

$$\int_0^S \frac{m}{IE}\,ds + V\int_0^S \frac{z+a}{IE}\,ds - H\int_0^S \frac{u+b}{IE}\,ds + [\mathcal{M} - Va + Hb]\int_0^S \frac{ds}{IE}$$

$$= \int_0^S \frac{m}{IE}\,ds + V\int_0^S \frac{z\,ds}{IE} + Va\int_0^S \frac{ds}{IE} - H\int_0^S \frac{u\,ds}{IE} - Hb\int_0^S \frac{ds}{IE}$$

$$+ \mathcal{M}\int_0^S \frac{ds}{IE} - Va\int_0^S \frac{ds}{IE} + Hb\int_0^S \frac{ds}{IE} = \int_0^S \frac{m\,ds}{IE} + \mathcal{M}\int_0^S \frac{ds}{IE} = 0.$$

Accordingly, $\mathcal{M}$ can be calculated from the simple formula:

$$\mathcal{M} = -\frac{\int_0^S \frac{m\,ds}{IE}}{\int_0^S \frac{ds}{IE}}. \tag{VIa}$$

Further, equation (IV$_2$) may be modified as follows:

ARCHES

$$\int_0^S \frac{m(u+b)}{IE} ds + V \int_0^S \frac{(z+a)(u+b)}{IE} ds + H \int_0^S \frac{(u+b)^2}{IE} ds$$

$$+ \int_0^S \frac{\mathcal{M}-aV+bH}{IE}(u+b) ds = 0.$$

It will be observed that

$$V \int_0^S \frac{(z+a)(u+b)}{IE} ds = V \int_0^S \frac{zu}{IE} ds + Va \int_0^S \frac{u\,ds}{IE} + Vb \int_0^S \frac{z}{IE} ds$$

$$+ Vab \int_0^S \frac{ds}{IE} = Vab \int_0^S \frac{ds}{IE}.$$

Also:

$$H \int_0^S \frac{(u+b)^2}{IE} ds = H \int_0^S \frac{u^2}{IE} ds + 2Hb \int_0^S \frac{u\,ds}{IE} + Hb^2 \int_0^S \frac{ds}{IE} = H \int_0^S \frac{u^2}{IE} ds + Hb^2 \int_0^S \frac{ds}{IE}.$$

And

$$\int_0^S \frac{\mathcal{M}-aV+bH}{IE}(u+b) ds = (\mathcal{M}-aV+bH)\left[\int_0^S \frac{u}{IE} ds + \int_0^S \frac{b}{IE} ds\right]$$

$$= \mathcal{M}b \int_0^S \frac{ds}{IE} - Vab \int_0^S \frac{ds}{IE} + Hb^2 \int_0^S \frac{ds}{IE}.$$

It follows that equation (IV$_2$) may be represented in the form:

$$\int_0^S \frac{mu}{IE} ds + b \int_0^S \frac{m}{IE} ds + Vab \int_0^S \frac{ds}{IE} - H \int_0^S \frac{u^2}{IE} ds - Hb^2 \int_0^S \frac{ds}{IE} + \mathcal{M}b \int_0^S \frac{ds}{IE}$$

$$- Vab \int_0^S \frac{ds}{IE} + Hb^2 \int_0^S \frac{ds}{IE} = \int_0^S \frac{mu}{IE} ds + b \int_0^S \frac{m}{IE} ds$$

$$- H \int_0^S \frac{z^2}{IE} ds + \mathcal{M}b \int_0^S \frac{ds}{IE} = 0. \tag{IV$_4$}$$

Multiplying equation (VIa) by b we find:

$$\mathcal{M}b \int_0^S \frac{ds}{IE} = -b \int_0^S \frac{m\,ds}{IE}.$$

Substituting this formula in equation (IV$_4$) we obtain:

$$\int_0^S \frac{mu}{IE} ds - H \int_0^S \frac{u^2}{IE} = 0.$$

So that

$$H = \frac{\int_0^S \frac{mu}{IE}\,ds}{\int_0^S \frac{u^2}{IE}\,ds}. \qquad (IVa)$$

By means of this formula the calculation of the thrust, H, is much simplified.

It can be shown, in the same manner, that the vertical reactions, V, can be found from the equation:

$$V = -\frac{\int_0^S \frac{mz}{IE}\,ds}{\int_0^S \frac{z^2}{IE}\,ds}. \qquad (Va)$$

Thus, we have obtained three equations by means of which the unknown values $\mathcal{M}$, V and H can be found.[1] In practical calculations it is preferable to use these equations, instead of equations (IV_3), (V_3) and (VI_3).

When $\mathcal{M}$, V and H are thus found, we obtain M from

$$M = \mathcal{M} - aV + bH.$$

Note that for circular arches of constant section we will have (see Fig. 12)

$$b = \frac{2r}{\pi} = 0.6366r \qquad \text{if } \theta = 180°$$

and

$$b = \frac{2\sqrt{2}\,r}{\pi} - \frac{r}{\sqrt{2}} = 0.1932r \qquad \text{if } \theta = 90°.$$

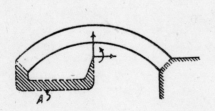

Fig. 11.

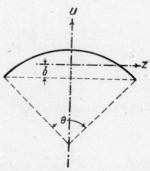

Fig. 12.

[1] As far as the author's knowledge goes, the analytical deduction of these three equations is original. During his lectures he has observed that this method of explanation was more easily grasped by the students, rather than the more usual demonstration including the imaginary cantilever A (see Fig. 11).

ARCHES

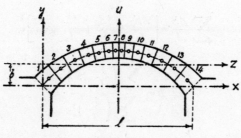

FIG. 13.

4. Solution of the Elastic Equations in Practice

Provided E is constant the results of the formulae (IVa), (Va), and (VIa) are independent of its numerical value. Consequently, in investigating the stability of a masonry arch, E is often assumed equal to unity and does not appear in the calculation at all, except in considering the effect of temperature and shrinkage. This point will be referred to again later on, in a special sub-section.

In calculating the unknown values of H, V and M, numerical or graphical solutions may alternatively be used. In some cases it may appear preferable to combine the two methods; for instance, compute graphically m, and then calculate all the other values arithmetically. We will first consider the numerical method.

(a) Arithmetical Solutions

The essential principle of this method consists in replacing the integral by a finite sum of small elements.

The calculation is made as follows: the arch is divided into a number of sections (see Fig. 13) and the values of y corresponding to the centroids of each section are figured out; then, $\int_0^S \frac{y\,ds}{I}$ is assumed to be approximately equal to the sum of these values, divided by the moments of inertia and multiplied by the lengths of the corresponding intervals, so that

$$\int_0^S \frac{y}{I}ds = \sum \left(\frac{y}{I}\,\Delta S\right) = \left[\frac{y_1}{I_1}\,\Delta S_1 + \frac{y_2}{I_2}\,\Delta S_2 + \frac{y_3}{I_3}\,\Delta S_3 + \ldots\right]$$

where y_1, y_2, y_3, etc., are the values of y corresponding to the centroids of the sections 1, 2, 3, etc., every centroid being supposed to coincide with the point on the axis of the arch at the half-length of the section (these points are marked by small circles in Fig. 13).

I_1, I_2, I_3, etc., are the moments of inertia of the cross-sections of the arch corresponding to the centroids of the sections, as determined above; the calculation being usually referred to one metre, or one foot, width of the arch perpendicular to the drawing; if D_1, D_2, D_3...etc., are meant to represent the thicknesses of the arch, measured at the centres of the sections 1, 2, 3, etc., then: $I_1 = \frac{D_1^3}{12}$; $I_2 = \frac{D_2^3}{12}$; $I_3 = \frac{D_3^3}{12}$, and so on.

ΔS_1, ΔS_2, ΔS_3, etc., are the lengths of the sections 1, 2, 3, etc., measured along the axis of the arch ring.

Under the same assumptions $\int_0^S \frac{ds}{I}$ is assumed to be equal to the sum:

$$\sum \frac{\Delta S}{I} = \left[\frac{\Delta S_1}{I_1} + \frac{\Delta S_2}{I_2} + \frac{\Delta S_3}{I_3} + \ldots\right].$$

Accordingly

$$b = \frac{\sum\left(\frac{y}{I}\Delta S\right)}{\sum\left(\frac{\Delta S}{I}\right)}.$$

For symmetrical arches, $a = \frac{l}{2}$ where l is the theoretical span of the arch, i.e. the distance between the centre points of the cross-sections at the springings (see Fig. 13).

When a and b are determined according to the formulae given above, the values u, z and m are figured out, for the centres of gravity of each section. Note that m is the bending moment calculated on the assumption that the left-hand end of the arch is free, while the right-hand end is fixed. In calculating m take the same width of the arch as is included in the calculation of I; i.e. either one metre or one foot (except in the case of very narrow arches, or arches with specific sections). When the arch is loaded by its own weight and by the weight of pedestrians or vehicles, m will be negative (according to the assumed direction of moments).

Then

$$\mathcal{M} = -\frac{\left[\frac{m_1}{I}\Delta S_1 + \frac{m_2}{I_2}\Delta S_2 + \frac{m_3}{I_3}\Delta S_3 + \ldots\right]}{\left[\frac{\Delta S_1}{I_1} + \frac{\Delta S_2}{I_2} + \frac{\Delta S_3}{I_3} + \ldots\right]}. \quad (VIb)$$

$$H = \frac{\left[\frac{m_1 u_1}{I_1}\Delta S_1 + \frac{m_2 u_2}{I_2}\Delta S_2 + \frac{m_3 u_3}{I_3}\Delta S_3 + \ldots\right]}{\left[\frac{u_1^2}{I_1}\Delta S_1 + \frac{u_2^2}{I_2}\Delta S_2 + \frac{u_3^2}{I_3}\Delta S_3 + \ldots\right]}. \quad (IVb)$$

$$V = -\frac{\left[\frac{m_1 z_1}{I_1}\Delta S_1 + \frac{m_2 z_2}{I_2}\Delta S_2 + \frac{m_3 z_3}{I_3}\Delta S_3 + \ldots\right]}{\left[\frac{z_1^2}{I_1}\Delta S_1 + \frac{z_2^2}{I_2}\Delta S_2 + \frac{z_3^2}{I_3}\Delta S_3 + \ldots\right]}. \quad (Vb)$$

An important simplification of the above solution has been suggested by Schönhofer[1]; namely, he proposes to make the length of each section inversely proportional to the cube of its depth, so that $\frac{\Delta S_n}{I_n}$ is constant for the whole length of the arch. Then, in calculating the integrals, the term $\frac{\Delta S_n}{I_n} = K$ can be taken out of the brackets; for instance:

$$\sum\left(\frac{y}{I}\Delta S\right) = [y_1 K + y_2 K + y_3 K + \ldots] = K[y_1 + y_2 + y_3 + \ldots].$$

Let the number of divisions be N; then

$$b = \frac{[y_1 + y_2 + y_3 + \ldots]}{N},$$

[1] Schönhofer, *Statische Untersuchung von Bogen- und Wölbtragwerken* ......(Berlin, 1908).

ARCHES

in other words, b, in this case, is the mean of all the values of y.

Similarly

$$\mathcal{M} = -\frac{[m_1 + m_2 + m_3 + \ldots]}{N} \tag{VIc}$$

i.e. $\mathcal{M}$ is the mean of all the values of m. Further:

$$H = \frac{[m_1 u_1 + m_2 u_2 + m_3 u_3 + \ldots]}{[u_1^2 + u_2^2 + u_3^2 + \ldots]} \tag{IVc}$$

$$V = -\frac{[m_1 z_1 + m_2 z_2 + m_3 z_3 + \ldots]}{[z_1^2 + z_2^2 + z_3^2 + \ldots]} \tag{Vc}$$

Note that in the case when the thickness of the arch is the same over all its length, the condition $\frac{\Delta S}{I} = $ const. is fulfilled by making all the divisions equal; this, however, is a particular case only, which, while it often occurs in the design of short spans, will almost never be met with in more important arches. In fact, it will usually be more convenient to increase the depth of the arch-ring, from the crown to the abutments, and in this case the length of Schönhofer's divisions must be made variable. The next problem, therefore, consists in dividing the centre-line of the arch into a number of short lengths, in such a manner as to satisfy the condition

$$\frac{\Delta S}{I} = \text{const.}$$

This is done as follows:

First the axis of the arch is developed into a straight line (for one-half of the length of the arch-ring in case of symmetrical layouts). Then the values of I are calculated for a number of points and two symmetrical curves are plotted, as shown in Fig. 14, using the calculated values of I as ordinates.

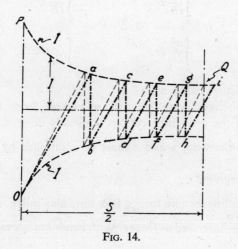

Fig. 14.

The line Oa is then drawn at a certain angle α to the line OP; this angle being chosen in the first computation tentatively, depending on the assumed number of divisions. A vertical line is thereafter traced through the point *a* until it meets in *b* the second curve of *I*. The operation is repeated in such a way that all the inclined lines *bc*, *de*, *fg* and *hi* are parallel to *Oa*, while *cd*, *ef* and *gh* are vertical; should the point *i* be found to coincide with *Q*, then the computation is completed, because the individual sections *TA*, *AB*, *CD*, etc., are then nearly proportional to the average *I* in every individual interval. In the general case, however, there will be a certain distance between *i* and *Q*. The computation must then be repeated, possibly several times, until *i* will fall exactly upon *Q*. This, however, takes less time than one might expect, particularly if the axes and the curves of *I* are traced in ink, before the preliminary polygons *Oabc*... are drawn in pencil.

(b) Graphical Solutions

The arithmetical calculation method outlined in the foregoing paragraphs supplies a perfectly simple and straightforward solution for the calculation of the elastic reactions, so long as we deal with *stationary* loadings; in fact, once the reactions are determined, the line of pressure is drawn (in exactly the same way as was shown earlier in Fig. 2), and the problem is thus solved.

It is, nevertheless, to be observed that in the case of *movable loads*, the application of the arithmetical solution is sometimes too laborious. It is then preferable to employ graphical computations, in applying in practice the equations furnished by the elastic theory.

There are two such methods: (*A*) The method of influence lines, which is common to all branches of modern bridge practice, and (*B*) a specific method, applicable to arches alone, which can best be described as the "elastic reaction-line method".

Both methods can be applied either precisely or in a simplified form taking advantage of ready-made tables and of a number of "short cuts" suggested by various authors.

A. Influence-lines Method. The standard computation is based on the fact that the influence lines for the factors

$$H = \frac{\int_0^S \frac{mu}{IE} ds}{\int_0^S \frac{u^2}{IE} ds} \quad \text{and} \quad V = -\frac{\int_0^S \frac{mz}{IE} ds}{\int_0^S \frac{z^2}{IE} ds}$$

can be represented by funicular polygons.

The ordinates *u* and abscissae *z* are then used as vectors, in plotting the respective stress diagrams. The division of the arch into sections is usually done according to Schönhofer's method $\left(\frac{\Delta S_n}{I_n} = \text{const.}\right)$. Otherwise, *u* and *z* must be multiplied by the individual values of $\frac{\Delta S_n}{I_n}$, before plotting the diagram.

In the same way, the influence line for $\mathcal{M}$ is a funicular polygon drawn for a stress diagram in which values of $\frac{\Delta S_n}{I_n}$ are used as forces. With Schönhofer's method all these forces are, however, equal.

The computation is illustrated in Fig. 15.

Let us consider the formulae (IV*c*) and (V*c*) (p. 17).

ARCHES

The denominators
$$u_1^2 + u_2^2 + u_3^2 + \cdots$$
and
$$z_1^2 + z_2^2 + z_3^3 + \cdots$$

are calculated once and for all, and do not appear in the computation (except in calculating the scale of the diagram), because they remain constant, whatever the position of the load on the arch.

On the other hand, the numerators
$$m_1 u_1 + m_2 u_2 + m_3 u_3 + \cdots$$
and
$$m_1 z_1 + m_2 z_2 + m_3 z_3 + \cdots$$

vary as the force moves from point to point on the arch, and it is this particular variable member of the formula which is given (to a certain scale) by the corresponding ordinates of the funicular polygon in question.

To explain this point attention is called to Fig. 16, which is supposed to represent a part of the graphical computation. The force "unity" is applied at the point x. In accordance with our original conventions (see p. 10) we will consider only that part of the arch which is located to the right of the point x, because all the moments, m, to the left of this point, must necessarily vanish; in fact, it will be remembered that in calculating m we consider only those of the external forces which are applied to the left of a section, and since in the present case there is one force only (which is applied at x) all the external moments to the left of the point x are axiomatically $=0$.

Let us now examine the moments m applied to the section between x and the right abutment. There are three sections in this part; remembering that in this computation the force is equal to unity, the corresponding moments are:

$$m_1 = r_1 \times 1.00$$
$$m_2 = r_2 \times 1.00$$
$$m_3 = r_3 \times 1.00.$$

It follows that for the point x the sum $\Sigma(mu)$ is equal to
$$r_1 u_1 + r_2 u_2 + r_3 u_3.$$

Consider, now, the corresponding ordinate, $f = ad$, of the funicular polygon. It is the sum $ab + bc + cd$. Since the triangle aob is similar to the triangle AOB, we will have

$$\frac{ab}{r_1} = \frac{u_1}{p}$$

from which
$$ab = u_1 r_1 / p.$$

In the same way, for bc and cd, we obtain
$$bc = u_2 r_2 / p$$
$$cd = u_3 r_2 / p.$$

It follows that the ordinate, f, of the polygon is
$$ad = \frac{u_1 r_1 + u_2 r_2 + u_3 r_3}{p} = \frac{\Sigma(mu)}{p}.$$

Table
(in metres)

1	2	3	4	5	6	7	8	9
	from diagram							
Point	Abscissa z	Original Ordinate	Final Ordinate $y-\Sigma y/n = u$	z^2	u^2	Arch thickness	I second moment	Constant $I/\Delta s$
1	0,575	3,920	+1,030	0,3306	1,0609	1,214	0,1440	0,1275
2	1,740	3,860	+0,970	3,0276	0,9409	1,243	0,1598	0,1330
3	2,970	3,775	+0,885	8,8209	0,7832	1,274	0,1720	0,1341
4	4,280	3,640	+0,750	18,3184	0,5625	1,307	0,1858	0,1355
5	5,680	3,440	+0,550	32,1489	0,3025	1,343	0,2020	0,1362
6	7,190	3,150	+0,260	51,6961	0,0676	1,381	0,2190	0,1370
7	8,820	2,780	-0,110	77,7924	0,0121	1,422	0,2395	0,1381
8	10,540	2,250	-0,640	111,0916	0,4096	1,466	0,2620	0,1393
9	12,400	1,540	-1,350	153,7600	1,8225	1,516	0,2880	0,1387
10	14,400	0,550	-2,340	207,3600	5,4756	1,570	0,3225	0,1375
Sums		28,905	0,005	664,346	11,437	—	—	1,3569

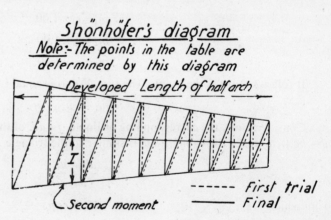

Shönhöfer's diagram
Note:—The points in the table are determined by this diagram

Developed Length of half arch

I Second moment

------ First trial
——— Final

FIG. 15.—Example of graphical

We can therefore write:

$$\Sigma(mu) = f \times p.$$

Q.E.D.

It goes without saying that a similar proof could also be given for the influence-line computation for V and $\mathcal{M}$.

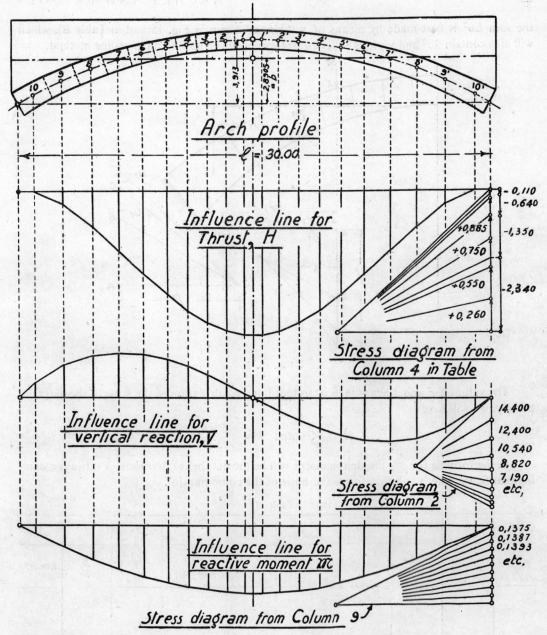

computation of influence lines.

In order to avoid an unfortunate but rather frequent mistake, attention is called to the fact that in the last formula the ordinate, f, of the funicular polygon is measured to the scale adopted for drawing the outline of the arch, whereas the polar distance, p, is measured to the scale employed in plotting the stress diagram.

To complete the analysis, it remains to divide the product fP by Σu^2. The calculation of the sum Σu^2 is best made by means of a table (as shown in Fig. 15 and in Table B), which will also contain Σz^2 and various other constants appearing in the influence-line method.

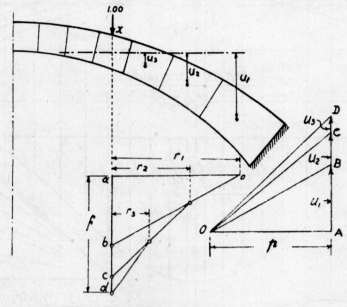

Fig. 16.

The values Σu^2 and Σz^2 are then introduced in the calculation of the scale of the diagram; namely, we will have

$$H = \frac{f \times p}{\Sigma u^2} \quad \text{and} \quad V = -\frac{f_1 \times p_1}{\Sigma z^2}.$$

In the formula for $\mathcal{M}$ the denominator will be the number of divisions, N (always assuming that Schönhofer's solution forms the basis of the computation).

Table B

1	2	3	4	5	6	7	8	9	10
Point	z	y	$u = y - b$	z^2	u^2	Arch thickness	I	$\frac{I}{\Delta S}$	Remarks
Totals	Σz	Σy	Σu	Σz^2	Σu^2				

ARCHES

In the above table the values u are calculated from

$$u = y - \frac{\Sigma y}{N}$$

and the values z from

$$z = x - \frac{l}{2}$$

where l is the theoretical span, i.e. the horizontal distance between the centres of skewbacks.

The rest of the computation in the table is obvious.

B. Elastic Reaction-curve Method. Imagine an arch subject to the effect of one force, P, only (see Fig. 17). Let the reactions at the right-hand and left-hand abutments be R and R_1 (as before). Under the effect of these three forces (P, R and R_1) the arch is in equilibrium. It follows that P, R and R_1 must intersect in one point, C.

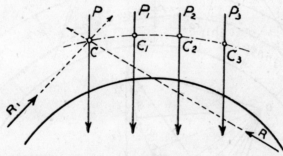

FIG. 17.

Suppose that the force P moves and occupies seriatim the positions P_1, P_2, P_3, etc. For each of these positions there must exist a point in which the two reactions and the force P intersect. Let these points be C_1, C_2, C_3 (see Fig. 17). We can then draw a line through these various points and obtain a curve, which will be referred to in these pages as the "elastic reaction-curve".

We shall begin the discussion by showing that, should such a line be drawn, the problem of stresses in the arch would thereby be indeed solved.

In fact, suppose that the loading of the arch is composed of a set of forces $P_1, P_2, \ldots P_8$ (see Fig. 18). Apart from the structural weight these forces are also supposed to include the live loads. (It will be observed that the loading on the left half of the arch in Fig. 18 is intentionally shown to be substantially heavier than that on the right side; this is done because unsymmetrical loading usually furnishes the critical conditions in such calculations.)

We shall now assume that, by some means or other, the elastic reaction-curve has been computed. We can then resolve each of the P forces into R' and R (left and right reactions) with a corresponding index. The vectors $(R')a$, $(R')b$, $(R')c\ldots$, etc., are then plotted in a stress diagram and a funicular polygon is drawn, using in this computation a certain pole O_1.

By means of this polygon we locate graphically the total reaction, R', at the left abutment (resultant of all the forces $(R')a$, $(R')b$, $(R')c$, etc.).

From the point D we will then plot all the forces P_1, P_2, P_3, etc. (downwards), and starting from the point A (in which R' meets P_1) draw another funicular polygon, taking O_2 as the new pole.

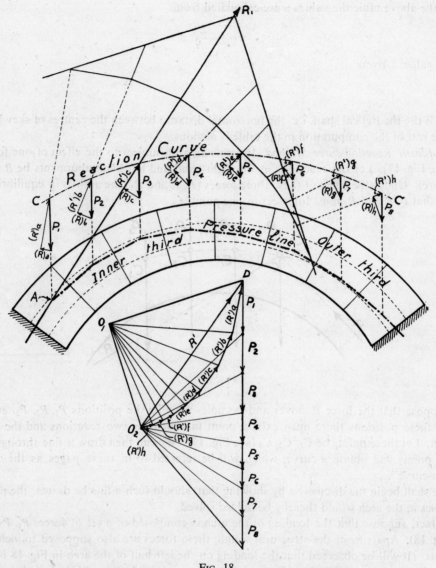

Fig. 18.

This polygon is, then, the line of pressure, and the problem is thus solved.

It remains to show how the reaction-curve can be computed.

There are two schools of thought in connection with this problem.

Some French arch builders, for instance Séjourné, adopt the precise solution, involving the computation of the ellipse of deformations, etc.[1] On the other hand, the German School, which includes Professor Landsberg,[2] Professor Mehrtens[3] and others, maintain that the

[1] P. Séjourné, *Grandes Voutes* (Bourges, 1913).
[2] Th. Landsberg, "Beitrag sur Theorie der gewolbe", *Zeitschr. d. Ver. deutscher Ing.* (1901).
[3] G.C. Mehrtens, *Vorlesungen über Ingenieur-Wissenschaften* (Leipzig, 1912).

ARCHES

influence-line method, described earlier, furnishes in itself a precise and exhaustive solution which does not call for further improvement, and that the second method (that of the elastic reaction-curve) should, therefore, be confined to the cases, only, in which an approximate but sufficiently accurate solution is aimed at.

With this object in view the German authors have suggested a number of computation methods, based on the elastic theory, which nevertheless make the tracing of the reaction-curve almost as easy as the graphical computation methods, belonging to the conventional solutions of the nineteenth century (three-points-method, etc.).

It will be clear, however, that the reaction-curves obtained in that manner cannot (and are not intended to) apply universally to all types and shapes of arches, but are intended to be used in each case for a certain, given type of arch layout.

(a) *Professor Landsberg's Solution*

This method is designed for the frequently employed type of arch with a parabolic axis (see Fig. 19), the arch-thickness increasing from crown to skewback according to the formula $D = D_0/\cos \phi$.

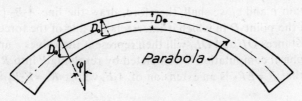

Fig. 19.

Professor Landsberg finds that the reaction-curve for this type of arch is a horizontal straight line, drawn at a distance equal to one-fifth of the theoretical rise, f, above the centroid of the section, at the crown (see Fig. 20).

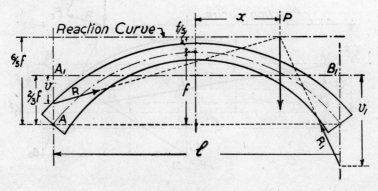

Fig. 20.

In addition to this, he gives the following method for finding the reactions R and R'.

A horizontal line A_1B_1 is drawn in the diagram, at a distance $2f/3$ above the line AB joining the centroids of the sections at the skewbacks.

The vertical distance v between A and the point in which the reaction R intersects the vertical line AA is determined from

$$v = \frac{8}{5} \times f \times \frac{l}{l+2x}$$

in which x is the distance between the force P and the vertical axis, as shown in Fig. 20.

On the other hand, for the reaction R_1 (at the skewback nearest to P) the corresponding vertical distance v_1 (see drawing) is found from

$$v_1 = \frac{8}{5} \times f \times \frac{l}{l-2x}.$$

Here again, instead of arithmetic, we can use a simple graphical solution; in fact the last two equations may be transcribed as follows:

$$v : \frac{18}{15} f = \frac{l}{2} : \left(\frac{l}{2} + x\right)$$

and

$$v_1 : \frac{18}{15} f = \frac{l}{2} : \left(\frac{l}{2} - x\right).$$

In order to obtain v and v_1 we shall, therefore, draw the line A_2B_2 (see Fig. 21) at $8f/15$ below A_1B_1 and join the point D in which it intersects the line of the force P, with A_1 and B_1, respectively. The distances OL and OL_1 will then represent the values v and v_1.

The purely graphical computation is completed by resolving P into R and R_1 as shown in the drawing. Note that $R = EE_2$ is an extension of A_3E, whilst $R_1 = E_2E_1$ is drawn parallel to B_3E.

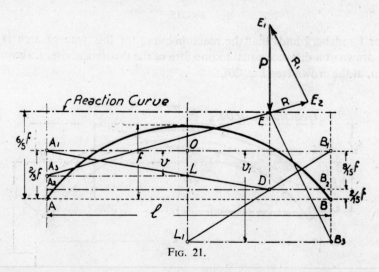

Fig. 21.

This operation is repeated for every force P and thereafter the polygons shown in Fig. 21 are drawn, and the problem solved.

(b) Professor Mehrtens' Solution

This applies to a larger variety of arch designs; as a matter of fact, the only specific condition qualifying the scope of application of this solution, is that the thickness of the arch-ring must increase from crown to skewback, so that we could approximately write:

ARCHES

$$I \cos \phi = I_0.$$

In this case Professor Mehrtens develops for the ordinates w of the elastic-reaction line the simple equation

$$w = \frac{2Pa^2b^2}{Hl^2}.$$

As shown in Fig. 22 the ordinates w are measured above the axis I–II, determined in the same manner as explained on p. 11 in connection with formulae (A).

It will be observed that in the equation for w the ratio P/H remains, so far, undetermined; in fact, this ratio must be found from specific considerations, characteristic of each particular case. For instance, for a parabola Professor Mehrtens finds:

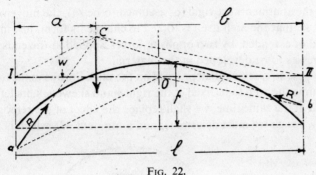

Fig. 22.

$$\frac{P}{H} = \frac{4fl^2}{15a^2b^2}.$$

Substituting this into the equation for w we obtain:

$$w = \frac{8}{15}f,$$

which is the same formula as that of Landsberg (straight line). Thus, Landsberg's result may possibly be considered as a particular case of the more general formula of Mehrtens.

In regard to the computation of the reaction, Mehrtens uses a simpler solution, which is due originally to Winkler. In fact, in order to find the points a and b in which the reactions intersect the verticals passing through the centroids of the skewbacks, it will suffice to join c with I and II, and then, draw Oa and Ob parallel to cI and cII, respectively. The rest of the computation is the same as before.

5. Effect of Temperature

Had the arch been entirely free to move in all directions, it would have expanded or contracted under the effect of ambient temperature variations, without creating any new stresses.

Thus, for instance, should temperature have risen by t degrees, the arch ABA_2B_2, which is shown hatched in Fig. 23, would have expanded into $A_1B_1A_3B_3$ (dotted outline), the sum of the distances $AA_1 + A_2B_3$ being equal to

$$\Delta l = \alpha t l,$$

where α is the coefficient of thermal expansion of the particular material of which the arch is supposed to be built.

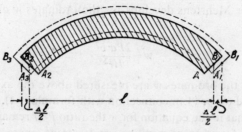

FIG. 23.

However, since the abutments are rigid (or assumed to be so), the movements at the skewbacks are prevented, so that the arch itself is forced to change its form, just in the same way as if it were compressed or extended, by two opposite forces applied at the ends. This, naturally, causes additional stresses to be developed in the material.

In order to find these stresses we shall proceed on the same lines as in the foregoing analysis, namely, we shall assume the arch to be fixed at the right-hand skewback, its left-hand end being left free to move; but at the same time, we shall replace the effect of the removed left abutment by a certain force H_t, which we must determine, and a moment M_t, which is also unknown.

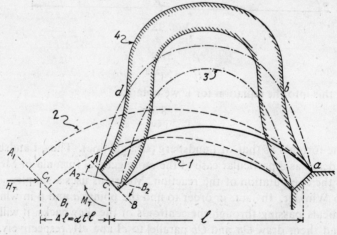

FIG. 24.

In order to explain the manner of action of this moment, attention is called to Fig. 24, in which the arch is represented as a thin, extremely elastic band, its thermal movements being intentionally exaggerated, so as to be able to trace separately the effect of several different mechanical actions; which in Nature occur all at the same time.

The full line (profile 1) represents the arch, as it is supposed to have been before the temperature began to change. The dotted profile (profile 2) gives the imaginary position of the arch, after the change of the temperature had taken place, but before the internal forces were developed. It will be observed that the centre point of the left skewback is shown to have travelled from c to c_1.

To bring the point back into its original position, a certain force H_t must be applied. This

ARCHES

force will cause the beam to be bent, as shown by the dot-and-dash lines (profile 3). It will be observed, however, that the force H_t acting alone, does not suffice to place the skewback into its first position (as it was originally, i.e. before the temperature rose), because owing to bending, the slope of the line AB has now changed, the new position being indicated in the drawing by A_2B_2. It follows that a moment M_T must also be applied in conjunction with the force H_T, in order to maintain the skewback in its original position. The effect of this moment will cause the shape of the arch to be altered, as shown by profile 4.

In order to avoid misunderstandings on the subject, attention is called again to the fact that the forces and moments, which are herein described as acting successively one after the other, are indeed developed in nature, from zero to their maximum value (and produce their effect upon the material) *simultaneously* with the rise of the temperature; so that the skewback remains always in its original position. The equations, however, will remain the same whatever the sequence of the different actions considered, that is to say, whether they occur concurrently or seriatim.

In considering the fourth profile (final), the following point will be clear: in regard to its deformation, the arch subjected to a temperature effect may be divided into three sections, namely: the two lateral parts ab and dc, in which the bending moment is positive (see arrow in Fig. 24 which indicates the assumed positive direction of moments) and the central part bd subject to a negative bending moment. It is also evident that b and d are the points of contraflexure, where the moment is zero.

Should the temperature fall, instead of rising, the moments will be inversed, but the location of the contraflexure points, b and d, will not change.

These general conclusions are of some importance to the designer, since they show that, if special bars are to be provided to take the temperature stresses (or, alternatively, if more material is to be added for the same object), this must be done near the crown and towards the springings, but not at quarter length of the span. Expansion joints (if any) must also be located in the zone of greatest thermal stresses, i.e. at the crown and stringings, but not at one third points of the span, as has been done in some earlier irrigation works in Egypt, for instance, the original Esna Barrage. Little wonder, therefore, that in spite of more than 200 joints provided in this barrage, over 30 cracks developed subsequently in the critical points.

Our next problem is: to find the formulae by means of which these temperature stresses could be calculated. Consider the condition in the section ab (see Fig. 25). The moment in this section is the sum of the moment M_t applied to the end of the beam and of the negative moment H_ty due to the force H_t. Accordingly the angular deflection of the element is

$$\frac{M_t - H_t y}{IE}.$$

Integrating this from 0 to S we get the total angular deflection

$$\int_0^S \frac{M_t - H_t y}{IE}\, ds. \qquad (a)$$

On the other hand, following the same line of reasoning as was adopted in the foregoing sub-section, in developing the "elastic" equations, we will find that the distortion of the element ab causes the point c to travel through the infinitely small distance

$$\frac{M_t - H_t y}{IE}\, y\, ds$$

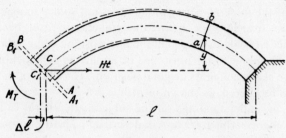

Fig. 25.

and that the total travel of the point c, due to the cumulative deflections of all the infinitely small elements of the arch, is expressed by

$$\int_0^S \frac{M_t - H_t y}{IE} \, y \, ds. \tag{b}$$

Now, from what was said before on the moment M_t and force H_t, it follows that their combined action is such as to neutralize completely the effect of thermal movements, *in so far as the section AB (arch springing) is concerned.* This means that:

(*a*) The slope of the section remains unchanged.
(*b*) The point c does not alter its position.

From (*a*) we have

$$\int_0^S \frac{M_t - H_t y}{IE} \, ds = 0. \tag{VII}$$

From (*b*)

$$\int_0^S \frac{M_t - H_t y}{IE} \, y \, ds = \Delta l = \alpha t l. \tag{VIII}$$

These two equations frame the required solution.

To facilitate the computation we may proceed in the same way as we did with the main elastic equations, namely introduce a new variable,

$$\mathcal{M}_t = M_t - b H_t,$$

and change the origin of co-ordinates in such a way that

$$u = y - b$$

where b is supposed to represent the same value as in sub-section 3, namely:

$$b = \frac{\int_0^S \dfrac{y \, ds}{IE}}{\int_0^S \dfrac{ds}{IE}}.$$

Substituting these formulae into equations (VII) and (VIII), we obtain

$$\int_0^S \frac{\mathcal{M}_t + bH_t - H_t y}{IE} ds = \int_0^S \frac{\mathcal{M}_t - H_t u}{IE} ds = 0 \qquad \text{(VIIa)}$$

$$\int_0^S \frac{\mathcal{M}_t + bH_t - H_t y}{IE} y\, ds = \int_0^S \frac{\mathcal{M}_t - H_t u}{IE} u\, ds + b \int_0^S \frac{\mathcal{M}_t - H_t u}{IE} ds = \alpha t l. \qquad \text{(VIIIa)}$$

Multiplying the first of these equations by b and subtracting it from the second:

$$\int_0^S \frac{\mathcal{M}_t - H_t u}{IE} u\, ds = \mathcal{M}_t \int_0^S \frac{u\, ds}{IE} - H_t \int_0^S \frac{u^2\, ds}{IE} = \alpha t l. \qquad \text{(VIIIb)}$$

On the other hand, equation (VIIa) may be written in the form:

$$\mathcal{M}_t \int_0^S \frac{ds}{IE} = H_t \int_0^S \frac{u\, ds}{IE}.$$

The term $\int_0^S \frac{u\, ds}{IE}$ has been shown, earlier, to be equal to zero. It follows therefore (since $H_t \neq 0$) that $\mathcal{M}_t = 0$, which means that equation (VIIIb) can be changed into

$$H_t \int_0^S \frac{u^2\, ds}{IE} = -\alpha t l$$

and therefore

$$H_t = -\frac{\alpha t l}{\int_0^S \frac{u^2\, ds}{IE}}. \qquad \text{(VIIIc)}$$

This is the formula which is commonly used for the calculation of temperature stresses in arches. The integral appearing in the denominator, in the right-hand term of this equation, is the same which is supposed to have been already calculated, when determining the value of the thrust due to the structural weight and the live loads (see formula (IVa)). The calculation of the temperature stresses is, therefore, done quite simply; namely, since $\mathcal{M}_t = 0$, the thermal bending moment in any section of the arch is equal to H_t multiplied by the corresponding ordinate u, as follows:

$$(M_T)_t = H_t u = -\frac{u \alpha t l}{\int_0^S \frac{u^2\, ds}{IE}}.$$

This formula throws more light on what was said earlier about the positive and negative thermal moments. In fact, it will be observed that the sign of the moment depends on u; namely, when u is positive $(M_0)_t$ is negative, and vice versa. Consequently, in the middle portion, delimited by the points a and b (see Fig. 26), where the centre-line of the arch-ring is located above the axis of z, the moment must necessarily be negative. Beyond these limits, that is to say, to the left of *a* and to the right of *b*, the ordinates *u* are negative and $(M_0)_t$ is, therefore, positive. It follows that *a* and *b* are the points of contraflexure, which divide the arch into three sections as explained earlier. Should the corresponding deflection be greatly exaggerated we would obtain the alignment shown in dotted lines in the drawing. It is very much the same as profile 4 in Fig. 24.

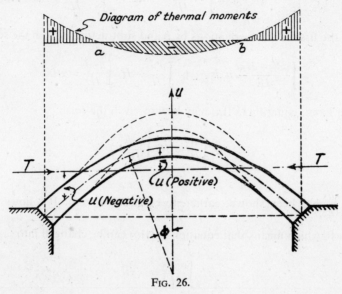

Fig. 26.

In the case if, instead of a rise, we have to consider a fall of temperature, the value of *t* will be negative, and all moments will be, therefore, reversed.

In calculating the thermal effect, one might add together the moments and thrusts calculated for the weight, live loads, and temperature, and then find the stresses corresponding to the total moment and total thrust thus calculated. This method of procedure, however, is very seldom followed, because it is found more convenient to calculate the temperature stresses separately, the latter being thereafter added to, or subtracted from, as the case might be, those due to the purely mechanical factors; the reason being: (*a*) that different permissible limits are often specified for the stress calculated with and without temperature, and (*b*) that it is, in certain cases, required to keep the line of pressure calculated for all mechanical loads, within the middle third of the arch; while a relatively light steel reinforcement is provided, at the same time, to take the thermal tensile stresses. Projects prepared according to this last principle, although they may appear rather conservative, are, nevertheless, quite popular, and are well adapted to the general spirit of irrigation design, which calls, possibly, for a higher factor of safety than is commonly adopted in other branches of civil engineering.

It follows therefore that, once H_t is known, the next step consists in calculating the corresponding stresses, namely:

$$\eta_t = \frac{H_t}{D \cos \phi} \pm \frac{12 u H_t}{D^3}.$$

Note that the thrust, H_t, appearing in this formula is supposed to be determined for one unit width of the arch.

ARCHES

The values of η_t, thus determined, are then added or subtracted from the stresses previously calculated.

As to the constants t, α and E to be included in these calculations, attention is called to the following. While some authors are inclined to consider in this case the external temperature variations, it is rightly objected by others, that a block of material, such as an arch, cannot be subject to the same range of variations as the air; the difference between maximum and minimum temperature being less in the first case than in the second. A total range of 30° to 40° centigrade is usually provided for, in specifications concerning arch design. We must therefore assume that t might be either positive or negative, with a maximum of 15° to 20° in both cases. It is obvious that, if special *thermal* reinforcement is to be provided, it must of necessity be symmetrical.

The usual value for α is 0.00001 per degree centigrade. In so far as arches are concerned, there is comparatively little difference of opinion about this coefficient (sometimes assumed 0.000008 per degree centigrade).[1] On the other hand, in regard to E, the range between maximum and minimum estimates and assumptions, made by different authors and under different conditions, is rather wider. The following figures given by Professor G.C. Mehrtens[2] may be quoted as an instance:

$E = 30,000 - 50,000$ kg per sq cm for brick arches
$70,000 - 100,000$,, ,, ,, ,, ,, stone arches
$350,000 - 400,000$,, ,, ,, ,, ,, concrete arches.

While the first two coefficients for bricks and stone appear to be possibly too low, the last figure in the table is rather on the high side, remembering that the ratio of Young's Moduli for steel and concrete, respectively, is supposed to vary from 8 for low stresses, to about 20 at breaking-point. As the modulus for steel is about $E = 2,100,000$ kg per sq cm, it might be more correct to adopt as E for concrete arches the figures given by Kelen,[3] namely:

$E = 100,000$ kg per sq cm in tension
$E = 200,000$,, ,, ,, ,, in compression.

To simplify the calculation, an average value of $E = 150,000$ kg per sq cm can perhaps be suggested.[4]

SECTION II. DESIGN PROCEDURE

1. The Problem

The elastic theory of arches, discussed in the preceding pages, places a powerful tool in

---

[1] The apparent value of α derived from observations of the thermal movements of the Aswan Dam was 2.3×10^{-6} per degree C, but as these were "restrained" movements, the author's analysis, based on the elastic principle, yielded 4.5×10^{-6} as the true value of this coefficient. This, however, was a particular case, which could not be applied in arch theory, because both materials and physical conditions are basically different.

[2] See *Vorlesungen über Ingenieur-Wissenschaften*, Vol. III, p. 30.

[3] See *Die Staumauern*, pp. 36 and 40.

[4] The value of E obtained by the author from the recorded movements of the Lock Wall at the Aswan Dam was 44,300 kg per sq cm, whereas for six other dams he found 45,000 kg per sq cm. In these figures, however, the elastic (or plastic) movement of the foundation is not accounted for, and this introduces a bias, the apparent value of the coefficient being too low. Note also that the E-values used by practising engineers in different branches of civil design (locks, bridges, dams, etc.) are not always the same, which testifies to the fact that the relevant theories are not perfect, including as they do a large proportion of empiricism.

the hands of the designer, but it does not suffice, alone, to solve the entire problem; for, though it supplies an excellent method for the calculation of stresses, this method can only be applied after (and not before) the arch-ring is designed.

Thus, had there been no other method available, the designer would have been obliged to proceed by trial and error; the section and alignment of the arch-ring being first guessed, and then verified by application of the elastic theory. In the absence of specific rules for the selection of the alignment and proper dimensioning of the arch, this method of procedure might have taken much time to complete.

It is therefore only natural that various approximate solutions have been suggested, as a preliminary step to be taken, before applying the elastic theory.

We do not refer here to the purely empirical rules for the design of arches. Such rules constitute a far too elementary solution, which may possibly be justified for very small spans, and light and current types of traffic, but do not embody the principles of rational analysis, which must necessarily control the design of more important works.

The problem may, however, be approached from another angle, viz. the arch, instead of being considered as a curved beam, may also be visualized as a crystallized funicular polygon, the bending effect being, thus, relegated to the class of undesirable secondary stresses, which should as far as possible be avoided (or reduced to a reasonable minimum).

The problem may then be stated as follows: could an arch be designed in such a manner, that its material was subject to compression alone, no bending stress being created at all?

Theoretically the reply is: yes; for consider the articulated system *abcdef* in Fig. 27 which carries the loads *ABCDEF*, and is supposed to follow the alignment of the funicular polygon drawn according to the stress-diagram (I).

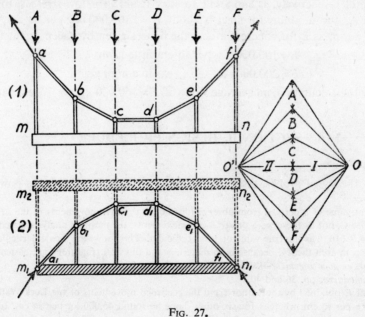

FIG. 27.

This system represents, evidently, a suspension bridge, the elements of which (*ab*, *cd*, etc.) are subject to extension only.

ARCHES

Suppose, now, that the pole of the stress diagram is shifted from O to O_1. The new stress diagram is described in the drawing as (II), and the corresponding funicular polygon is $a_1b_1c_1d_1e_1f_1$.

It is clear that the system $a_1b_1c_1d_1e_1f_1$ represents, schematically, an arch. Assuming the articulations in the points $a_1b_1c_1d_1e_1$ and f_1 to be frictionless, the elements a_1b_1, b_1c_1, c_1d_1, etc., are all subject to compression alone. Thus, an arch built on this principle would be free from bending stresses, and would therefore constitute the most economical design.

There is, however, a serious objection against applying this method in practice; in fact, in Nature, the loads $ABCD$, etc., can never remain constant, and a slight alteration in their value, or position, will suffice to destroy the equilibrium of the articulated system.

It is therefore clear that some means must be provided to ensure the rigidity of such an arch. Two methods can be used. We may either provide an additional member (stiffening truss) which will carry the additional stresses due to the alteration of the loading, and will thus prevent excessive deformation of the arch; or we can make the arch itself sufficiently stiff to ensure its rigidity.

Let us begin with the first alternative. The girder mn in the same figure is supposed to illustrate the classical solution of a stiffening truss of a suspension bridge. That such trusses can also be used in combination with flexible arches is, possibly, a less well known contingency; and yet, this solution has been proposed and occasionally used in practice. The stiffening girder can, in fact, be placed either below the arch (as shown in the drawing by m_1n_1) or above the arch (as represented by m_2n_2). In the first case we have the design which has been proposed by Langer[1] as early as 1871 (see Fig. 28), while the second case (stiffening girder above the

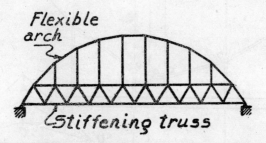

Fig. 28.

elastic arch) illustrates a principle embodied in some modern bridge designs[2] executed in Switzerland (see Figs. 29 and 30).

In both cases the arch is provided with enough material to carry the compressive stress imposed on it, but its rigidity is very low (as compared to that of the stiffening member) and it is, therefore, subject to no bending stress of any importance.

These, however, are exceptional solutions, which are quoted here only to make the general explanation of the subject more clear. The frequent, current solution is to make the arch itself

---

[1] Langer, "Festigkeitstheorie der Bruckenträger", *Technische Blaëtter* (1871).—Actually applied in the bridge over the Mur, in Graz, and in some modern Japanese bridges.

[2] See: "Die Linienverlegung der Rhät. Bahn in Klosters", by the Swiss Govt. Railways, published in *Schweiz. Bauzeitung*, Zürich, 20 Dec. 1930. Also: Prof. Dr. M. Roš, "Neuere Schweizertsche Eisenbeton-Brückentypen", same periodical, Vol. 90, Oct. 1927.

capable of resisting the bending effect, and the elastic theory is then applied to calculate its intensity.

The order of procedure will then be as follows: The arch is designed in such a way that its axis coincides with the funicular polygon drawn for all the dead load and for half the live load (average conditions). The cases of an entirely free arch, of the load over half-the-span and of the total load, constitute the extreme conditions, leading to the greatest departures from the original polygon. In plain concrete and masonry arches, which are supposed to carry no tensile stresses, all these polygons must lie within the middle third or the arch-ring. For reinforced concrete and steel arches this criterion is not of such importance, but here again, economy is attained if the tensile stress is reduced to a reasonable minimum.

In addition to that, the concept of an arch as a funicular polygon, or as an inverted suspension bridge, allows us to draw significant conclusions in regard to its general shape. In fact if, instead of a uniformly spaced loading, such as in Fig. 27, we consider a set of unequally distributed forces (see Fig. 31) it will at once be clear that the curvature of the polygon increases under the load concentrations. It follows that a suspension bridge designed to carry the loads A, B, C, D, E and F in Fig. 31, must be shaped as shown by *abcdef*; it is also clear that in the case of an arch working under similar conditions, the layout must be as shown by $a_1b_1c_1d_1e_1f_1$.

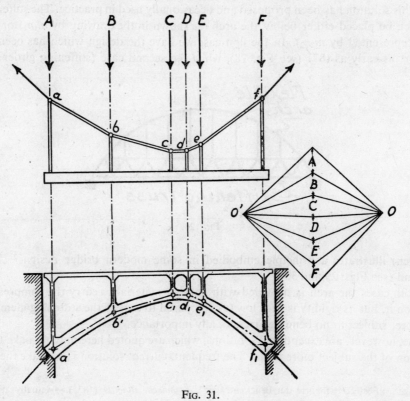

FIG. 31.

It will be realized that the irregularity of the loading shown in this drawing is intentionally exaggerated, which is done in order to explain more clearly the fundamental principles of arch

FIG 29.—Modern Swiss bridge design. Stiffening girders above elastic arch.

Fig. 30.—Modern Swiss bridge design.

ARCHES

design. In practice, concentrations of dead load only will be a significant factor for the correct choice of alignment of the arch axis; because live loads are an essentially movable element of the stress analysis, which has, therefore, no bearing on the permanent layout of the arch ring.

In regard, however, to the dead loads we can now state a rather important design principle; namely, that the elliptic arch (see Fig. 32 (a)) is more suitable for the filled spandrel arch, because the curvature of the arch increases under the heavy spandrels; whereas the open spandrel type must of necessity call for a parabolic alignment, as shown in Fig. 32 (b). In order to compare the two alignments, they are superposed in Fig. 32 (c).

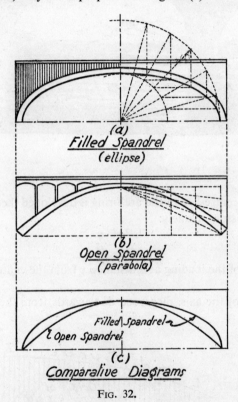

Fig. 32.

The elliptic form, however, is now scarcely ever used, being, as it is, replaced by the rationally developed parabola of the fourth degree, which is almost of the same shape.

To conclude this discussion, attention is called to an instructive instance of an early engineering design, represented in Fig. 33.[1]

In spite of the fact that this bridge was built in the twelfth century, its designer had a subconscious feeling of the true stress distribution, for without the heavy masonry gateway, placed at its middle-point, the bridge could have never been stable; in fact, at this point, the axis of the arch forms a sharp bend, and therefore, had there been no concentrated loading over it, no funicular polygon could have ever been made to pass within the thickness of the arch.

In the following study we explain Professor Melan's method, developed from the equation

---

[1] Hannibal's bridge over the Llobregat near Martorella, in the province of Barcelona (Spain).

of the funicular polygon for a filled spandrel arch, and we give two other methods for the calculation of the axis of the arch ring.

Fig. 33.

2. Melan's Method[1]

The equation of the centre-line of the arch-ring is developed theoretically from the general formula of the line of pressure, namely:

$$\frac{d^2y}{dx^2} = \frac{q}{H}$$

where q is the intensity of the loading at a distance x from the centre,

H is the thrust,

and y is the ordinate of the axis, measured downwards from the horizontal line II–II (see Fig. 34).

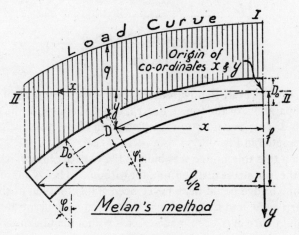

Fig. 34.

[1] See *Der Brückenbau* by Dr.-Ing. h. c. Yoseph Melan, Vol. II, 3rd ed., 1924.

ARCHES

It is understood that q includes all the dead loads and one-half of the equivalent distributed loading representing the live loads.

The final form of the equation is

$$y = \frac{1}{C}\left[1 + (C-1)\frac{4x^2}{l^2}\right]\frac{4x^2}{l^2} f.$$

In this formula:

$$C = \frac{1}{2}\left[\frac{q_0+q_2}{q_0} + \sqrt{\left(\frac{q_0+q_2}{q_0}\right)^2 + \frac{2}{3}\frac{\gamma_1 f}{q_0}}\right]$$

$q_0 =$ the value of q at the crown
$q_1 =$ the value of q at the skewback
$q_2 = 1/6\,(q_1 - q_0 - \gamma_1 f)$
$\gamma_1 =$ specific weight of the backing
$f =$ the rise
$l =$ the span

To facilitate the application of the formulae a table is prepared which gives C as a function of q_1/q_0 and $\gamma_1 f/q_0$.

Table giving the calculated values of C

q_1/q_0 \ $\gamma_1 f/q_0$	=0	1	2	3	4	5	6	7	8
1.0	1.000	1.000	—	—	—	—	—	—	—
1.5	1.083	1.072	—	—	—	—	—	—	—
2.0	1.167	1.145	1.129	1.115	—	—	—	—	—
2.5	1.250	1.220	1.195	1.175	—	—	—	—	—
3.0	1.333	1.295	1.284	1.237	1.215	1.196	—	—	—
3.5	1.417	1.371	1.333	1.301	1.274	1.250	—	—	—
4.0	1.500	1.448	1.404	1.366	1.333	1.306	1.282	1.260	—
4.5	1.583	1.526	1.476	1.432	1.395	1.364	1.331	1.309	—
5.0	1.667	1.604	1.549	1.500	1.457	1.420	1.387	1.339	1.333
5.5	1.760	1.682	1.622	1.569	1.521	1.479	1.456	1.410	1.382
6.0	1.833	1.761	1.696	1.639	1.587	1.541	1.500	1.464	1.431
6.5	1.917	1.840	1.771	1.709	1.653	1.603	1.558	1.518	1.483
7.0	2.000	1.920	1.847	1.781	1.721	1.666	1.618	1.574	1.535
7.5	2.083	2.000	1.923	1.853	1.789	1.731	1.679	1.632	1.589
8.0	2.176	2.080	2.000	1.926	1.859	1.797	1.741	1.690	1.644
8.5	2.250	2.161	2.077	2.000	1.929	1.864	1.804	1.750	1.706
9.0	2.333	2.241	2.155	2.074	2.000	1.931	1.869	1.811	1.758

Differentiating the formula for y and substituting $\frac{l}{2}$ for x, we obtain the tangent of the angle ϕ_0 (see Fig. 34) as follows:

$$\tan \phi_0 = \frac{4f}{l} \times \frac{2C-1}{C}.$$

The thickness of the ring is made variable conforming to the equation:

$$D = \frac{D_0}{\cos \phi}$$

where D_0 is the thickness at the crown.

For D_0 the following formulae are quoted:

Authors	Equations	
Perronet	$D_0 = 0.33 + 0.035l$	too large
Dejardin	$D_0 = 0.3 + 0.045l$	
Desnoyers	$D_0 = 0.15 + 0.175\sqrt{l}$	
Rankine	$D_0 = 0.191\sqrt{r}$	
Séjourné	$D_0 = \alpha(1 + \sqrt{l})\mu,$	

where, for flat elliptic arches,

$$\mu = \frac{4}{3 + 2n}$$

and for segmental arches $\quad \mu = \frac{4}{3}(1 - n + n^2);$

also $\quad n = \dfrac{f}{l}$

and $\quad \alpha = 0.15$ for road bridges
,, $= 0.19$,, railway bridges.

Heinzerling . . . . $D_0 = 0.4 + 0.025r$ for dressed stone
$D_0 = 0.4 + 0.028r$,, bricks
$D_0 = 0.42 + 0.032r$,, rubble

Schwarz . . . . $D_0 = 0.2 + \dfrac{1}{21} \times \dfrac{lQ}{S}$

where Q is the weight of one half of the arch with the corresponding live load, and S the permissible stress.

With regard to these formulae, it should be understood that Professor Melan uses them not directly, but as a means for finding q_0 in his own equation, which is:

$$D_0 = \frac{(\gamma_1 \mu + p) \rho_0}{S_0 + \gamma \rho_0} = \frac{(\gamma_1 \mu + p) r_0}{S_0 - \gamma r_0 - q_0},$$

where γ is the specific weight of the arch material
γ_1 ,, ,, ,, ,, of the backing
μ ,, ,, height of filling above the crown
ρ_0 ,, ,, radius of axis at crown
r_0 ,, ,, ,, of intrados at crown
S_0 ,, ,, permissible stress in the material in the central section at the crown.

The value of ρ_0 to be included in this formula can be found from

$$H_1 = q_0 \rho_0$$

where H_1 is the approximate value of the thrust, namely:

$$H_1 = C \times \frac{1}{8} \times \frac{q_0 l^2}{f}$$

which gives $\rho_0 = C \times \dfrac{l^2}{8f}.$

For short span arches this formula will be found to give rather too low values of ρ_0.

Another formula for D_0 is also mentioned by Melan. It is derived from considerations concerning the stability of a section taken midway between crown and skewback. For the arch thickness at crown it gives:

$$D_0 = \frac{1}{2}\frac{W\rho_0}{S-\gamma\rho_0} + \sqrt{\frac{1}{4}\left(\frac{W\rho_0}{S-\gamma\rho_0}\right)^2 + \frac{0.06 p_1 l^2 (\cos\phi_m)^2}{S-\gamma\rho_0}}$$

in which $W = \mu_0\gamma_1 - \frac{1}{2C}p_1$

p_1 = equivalent uniform loading for half the span;
ϕ_m = angle between the vertical and the normal section of the ring, at one-quarter of the span; for flat arches it will suffice to assume that

$$(\cos\phi_m)^2 = \frac{l^2}{l^2 + 4f^2}.$$

While for the larger part of its length the thickness of the ring is determined by $D = D_0/\cos\phi$, it is, nevertheless, recommended to add extra material in the vicinity of the springings, to account for local temperature stresses.

The two formulae for D_0 which are quoted above, are based on the maximum compressive stress. For it is assumed that the resultant falls automatically within the limits of the middle third of the arch-ring, in the case when the centre-line is designed to coincide (in so far as possible) with the pressure line. It should be understood, however, that for small spans it may be necessary to use deeper sections, so as to satisfy this last condition as such; namely, considering again the section at quarter-span distance from the crown, the "no tension" criterion leads to the following approximate equation:

$$D_0 \geq -\frac{W}{2\gamma} + \sqrt{\frac{W^2}{4\gamma^2} + \frac{p_1 f}{2\gamma C}(\cos\phi_m)^2}.$$

More accurate results are said to be arrived at, in using Professor Melan's method, if f is replaced by

$$f_1 = f\left(1 + \theta\frac{I_0}{Af^2}\right),$$

the coefficient θ being a function of the ratio

$$\delta = \frac{I_0}{I_1 \cos\phi_1}$$

where I_0 is the moment of inertia of the section at the crown
I_1 ,, ,, same moment at the skewback
A ,, ,, area of the section of the arch
ϕ_1 ,, ,, angle between skewback and vertical.
The function θ is tabulated below:

δ	θ	δ	θ	δ	θ
0	51.43	0.125	32.44	0.6	15.32
0.02	46.80	0.15	30.36	0.7	13.90
0.04	43.01	0.2	27.00	0.8	12.91
0.05	41.36	0.3	23.34	0.9	12.01
0.075	37.81	0.4	19.24	1.00	11.25
0.1	34.89	0.5	17.01		

The original publication also contains formulae for the bending moments at crown and springings, for the effect of temperature, for the live loads, and many other design factors, which will repay study.

3. Kögler's Method[1]

This author has calculated several tables, which may be used with advantage for the preliminary design of the arch-ring.

In the first instance, attention is called to the table for the calculation of the curve representing the axis of the arch-ring, should the latter be conceived as a pressure line, for a system of loads. The table (see p. 43) gives the values of y/f corresponding to various values of $\frac{2x}{l}$.

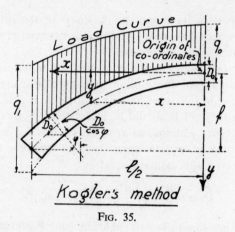

FIG. 35.

In order to use this table we must first determine the characteristic load-coefficient K, which depends on the ratio of the loadings q_0 and q_1 at the crown and skewback, respectively; namely, as follows (see Fig. 35):

$$K = \frac{q_1 - q_0}{6q_0}.$$

The depth of the ring is controlled, as before, by the equation:

$$D = \frac{D_0}{\cos \phi}.$$

The thrust (due to structural weight alone, or to structural weight plus a uniformly distributed loading) is found from:

$$H = abql^2/f,$$

in which the coefficients a and b are to be abstracted from the following table:

K	0.0	0.1	0.2	0.3	0.4	0.5	0.6	0.7	0.8	0.9	1.0	1.1	1.2
a	0.560	0.365	0.311	0.274	0.250	0.233	0.219	0.206	0.195	0.186	0.178	0.172	0.166
b	0.250	0.230	0.215	0.205	0.196	0.189	0.183	0.177	0.173	0.169	0.166	0.163	0.161

[1] See Fr. Kögler, *Vereinfachte Berechnung eingespannter Gewölbe* (Berlin, 1913).

ARCHES

Kögler's Table

giving values of $\frac{y}{f}$ as functions of $\frac{2x}{l}$ and K

$\frac{2x}{l}$ \ K	0.0	0.1	0.2	0.3	0.4	0.5	0.6
0	0	0	0	0	0	0	0
0.1	0.0100	0.0072	0.0064	0.0060	0.0057	0.0054	0.0051
0.2	0.0400	0.0352	0.0320	0.0300	0.0285	0.0271	0.0259
0.3	0.0900	0.0810	0.0752	0.0710	0.0674	0.0643	0.0617
0.4	0.1600	0.1463	0.1369	0.1297	0.1237	0.1186	0.1143
0.5	0.2500	0.2308	0.2184	0.2082	0.1995	0.1921	0.1856
0.6	0.3600	0.3367	0.3213	0.3088	0.2975	0.2876	0.2794
0.7	0.4900	0.4649	0.4476	0.4328	0.4206	0.4097	0.4002
0.8	0.6400	0.6167	0.5999	0.5860	0.5739	0.5633	0.5539
0.9	0.8100	0.7936	0.7813	0.7713	0.7627	0.7550	0.7481
0.975	0.9506	0.9459	0.9420	0.9389	0.9361	0.9337	0.9315
1.00	1.0000	1.0000	1.0000	1.0000	1.0000	1.0000	1.0000

$\frac{2x}{l}$ \ K	0.7	0.8	0.9	1.0	1.1	1.2
0	0	0	0	0	0	0
0.1	0.0040	0.0047	0.0046	0.0045	0.0044	0.0043
0.2	0.0249	0.0241	0.0234	0.0328	0.0223	0.0217
0.3	0.0596	0.0576	0.0560	0.0544	0.0531	0.0518
0.4	0.1106	0.1071	0.1040	0.1013	0.0990	0.0967
0.5	0.1803	0.1751	0.1708	0.1667	0.1628	0.1593
0.6	0.2722	0.2658	0.2600	0.2549	0.2496	0.2447
0.7	0.3920	0.3844	0.3777	0.3718	0.3658	0.3604
0.8	0.5456	0.5382	0.5318	0.5256	0.5202	0.5151
0.9	0.7420	0.7364	0.7314	0.7269	0.7226	0.7184
0.975	0.9295	0.9276	0.9258	0.9245	0.9231	0.9220
1.00	1.0000	1.0000	1.0000	1.0000	1.0000	1.0000

For the stress at the crown the following formula may be applied:

$$\eta = \frac{H}{A}\left(1 \pm \Delta \frac{D_0}{f}\right).$$

Here again a table is given which contains the values of the factor Δ as a function of the ratio $\frac{f}{l}$:

f/l	1/4	1/5	1/6	1/8	1/10	1/12
Δ	2.57	2.20	2.10	1.99	1.95	1.92

The stress due to temperature is found from

$$\eta_t = \pm \Delta \frac{D_0}{f} E\alpha t^5.$$

In the quoted formulae the live load, if any, is assumed to be distributed uniformly over the entire length of the span, but an approximate solution is also given, which may help in

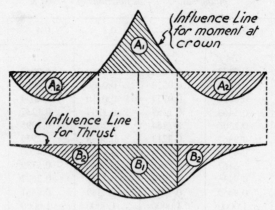

Fig. 36.

finding the effect of partial loading, occupying the most disadvantageous position; namely, with reference to Fig. 36, it will be clear that the maximum positive and negative values of the bending moment, at the crown, will be as follows:

Case I. Arch axis designed as pressure line for structural weight alone:

$$+M = (\text{area } A_1) P_1$$
$$-M = -2 (\text{area } A_2) P_2$$

In these formulae, A_1 and A_2 are, respectively, the positive and negative areas of the bending-moment influence-line diagrams, as shown in Fig. 36, and as tabulated below; whilst P_1 and P_2 are the uniformly distributed live loads corresponding to these areas (per unit length of the span).

f/l	K	0.0	0.2	0.4	0.6	0.8	1.0	1.2
1/4	A_1	4.87	5.51	6.03	6.47	6.80	7.08	7.33
	$2A_2$	4.51	3.97	3.57	3.29	3.09	2.93	2.80
1/5	A_1	5.18	5.80	6.30	6.75	7.13	7.50	
	$2A_2$	4.63	4.11	3.71	3.42	3.19	3.03	
1/6	A_1	5.38	5.96	6.49	6.98	7.48		
	$2A_2$	4.75	4.23	3.83	3.52	3.28		
1/8	A_1	5.60	6.11	6.69	7.32			
	$2A_2$	4.88	4.35	3.94	3.53			
1/10	A_1	5.75	6.19	6.88				
	$2A_2$	4.95	4.42	4.00				
1/12	A_1	5.85	6.28	7.01				
	$2A_2$	4.98	4.45	4.03				

ARCHES

Case II. Arch axis designed as pressure line for structural weight and half of the equally-distributed live load:

$$+M = +(\text{area } A_1) P_1 - (\text{area } A_1 - 2 \text{ areas } A_2) \frac{P}{2}$$

$$-M = -2 (\text{area } A_2) P_2 - (\text{area } A_1 - 2 \text{ areas } A_2) \frac{P}{2}$$

in which $P/2$ is the live load included in the calculation of the axis.

The areas A_1 and A_2 can be found from the table on p. 44, by multiplying the numbers appearing therein, by l^2 divided by 1000.

Note that a positive moment produces tension at the intrados and compression at the extrados.

To calculate the stress we must also know the thrust. This can be obtained from the second diagram represented in Fig. 36, namely:

For positive M . . . . . . $H = B_1 P_1$
For negative M . . . . . . $H = 2 B_2 P_2$.

The values B_1 and B_2 can be obtained by multiplying the figures listed in the following table by the factor l^2/f.

f/l \ K		0.2	0.4	0.6	0.8	1.0	1.2
1/4	$2B_2$	0.650	0.595	0.581	0.573	0.560	0.555
	B_1	0.643	0.730	0.770	0.802	0.834	0.859
1/5	$2B_2$	0.647	0.595	0.581	0.573	0.560	
	B_1	0.643	0.725	0.762	0.791	0.822	
1/6	$2B_2$	0.640	0.595	0.581	0.573		
	B_1	0.644	0.721	0.758	0.783		
1/8	$2B_2$	0.626	0.595	0.581			
	B_1	0.657	0.715	0.749			
1/10	$2B_2$	0.623	0.595				
	B_1	0.661	0.712				
1/12	$2B_2$	0.622	0.595				
	B_1	0.660	0.711				

4. Cochrane's Method[1]

This method was derived from the analysis of a number of complete designs, some of which were prepared "ad hoc" by its author, whilst others were abstracted from examples quoted in technical literature. The solution can therefore be considered as semi-empirical.[2]

[1] See "Design of Symmetrical Hingeless Concrete Arches" by V.H. Cochrane. Paper read before the Engineers Society of Western Pennsylvania, in 1916.

[2] This explains why it is that, whilst Melan's and Kögler's methods are in close agreement, the results of Cochrane's method are numerically different.

Let f/l be referred to as the "rise ratio" and denoted by the letter λ. The equation of the axis of the arch and the angle ϕ_1 between skewback and vertical will then be as follows:

Case I—Open Spandrel Arches

$$y = \frac{8\lambda l}{6+5\lambda}\left[3\left(\frac{x}{l}\right)^2 + 10\lambda\left(\frac{x}{l}\right)^4\right]$$

and

$$\tan\phi_1 = \frac{8\lambda}{6+5\lambda}(3+5\lambda).$$

Case II—Filled Spandrel Arches

$$y = \frac{4\lambda l}{1+3\lambda}\left[\left(\frac{x}{l}\right)^2 + 24\lambda\left(\frac{x}{l}\right)^5\right]$$

and

$$\tan\phi_1 = \frac{4\lambda}{1+3\lambda}(1+7.5\lambda).$$

In examining these formulae it will be observed that the effect of the term containing the higher powers of $\frac{x}{l}$ is more pronounced for the case of the filled spandrel, the ratio of loads at springing and crown, respectively, being greater in this case than for an open spandrel. This is in full agreement with the theory of the line of pressure, explained earlier (see Fig. 32).

In regard to the law controlling the variation of the depth of the arch-ring, Mr. Cochrane does not follow the standard equation $D = \frac{D_0}{\cos\phi}$, but uses the table reproduced below. The figures appearing in the first column give the ratios obtained by dividing the distances S_x measured along the axis of the ring from the crown to any given section, by $\frac{S_0}{2}$, which is the length of one-half of that axis. We must first find the thickness at the crown (employing for that purpose either an empirical formula or using any other approximate method) and assume a certain value of the ratio of thicknesses at skewback and crown, respectively. In regard to

Table 1. Ratio of Arch Thickness at Given Points to Crown Thickness

$\frac{2S_x}{S_0}$ \ $\frac{D_1}{D_0}$	1.5	1.75	2.0	2.25	2.50	2.75	3.0	3.25
0	1.000	1.000	1.000	1.000	1.000	1.000	1.000	1.000
0.05	1.007	1.006	1.005	1.004	1.003	1.002	1.001	1.000
0.15	1.021	1.018	1.018	1.015	1.012	1.009	1.006	1.000
0.25	1.035	1.030	1.025	1.020	1.015	1.010	1.005	1.000
0.35	1.049	1.042	1.035	1.028	1.023	1.021	1.023	1.030
0.45	1.063	1.054	1.048	1.048	1.057	1.070	1.083	1.101
0.55	1.077	1.072	1.085	1.105	1.133	1.163	1.193	1.231
0.65	1.095	1.125	1.168	1.215	1.268	1.328	1.385	1.455
0.75	1.145	1.223	1.311	1.403	1.508	1.625	1.737	1.865
0.85	1.245	1.393	1.547	1.700	1.862	2.025	2.185	2.355
0.95	1.406	1.621	1.827	2.055	2.277	2.495	2.709	2.932
1.00	1.500	1.750	2.000	2.250	2.500	2.750	3.000	3.250

ARCHES

the latter point it is suggested that for open spandrel arches D_1/D_0 might be taken from 1.5 to 2.5, while for the filled spandrel type, values of this ratio varying from 2 to 3.5 will be found convenient. It should also be remembered, in selecting the ratio in question, that higher values correspond to flat arches, whereas in the case of a relatively large rise, the thickness of the arch-ring should be made more uniform.

To help in calculating the length of the arch axis, when f and l are known, the following table is given:

Table 2. *Lengths of the Half Arch Axis $\frac{S_0}{2}$, in Terms of the Span Length l*

Values of rise-ratio	0.10	0.15	0.20	0.25	0.30
Open-spandrel arches	0.513	0.529	0.551	0.577	0.607
Filled-spandrel arches	0.515	0.534	0.559	—	—

To facilitate the verification of the design by means of the elastic method, it will be useful to refer to the figures given in Tables 3 and 4. They are both intended to help in calculating numerically the integrals in the basic formulae of the elastic method. In fact, Table 3 gives the constant ratio $\frac{\Delta s}{I} = K$ which is necessary for calculating arithmetically the formulae (VI$_b$), (IV$_b$) and (V$_b$) on page 16, whereas Table 4 replaces Schönhofer's computation in Fig. 14. In all cases the number of divisions in the half-arch is taken to be ten.

Table 3

Ratio $\frac{D_1}{D_0}$	Value of $\frac{\Delta s}{I}$	Ratio $\frac{D_1}{D_0}$	Value of $\frac{\Delta s}{I}$
1.5	0.0769 $S/2$ divided by I_0	2.50	0.0647 $S/2$ divided by I_0
1.75	0.0732 ,, ,, ,, ,,	2.75	0.0626 ,, ,, ,, ,,
2.00	0.0699 ,, ,, ,, ,,	3.00	0.0606 ,, ,, ,, ,,
2.25	0.0672 ,, ,, ,, ,,	3.25	0.0586 ,, ,, ,, ,,

Table 4. *Values of $\frac{2S_x}{S_0}$ of the Centres of Divisions having a Constant Ratio $\frac{\Delta S}{I}$ throughout*

No. of Division from Crown \ $\frac{D_1}{D_0}$	1.5	1.75	2.0	2.25	2.50	2.75	3.0	3.25
1	0.039	0.037	0.036	0.034	0.033	0.032	0.031	0.030
2	0.119	0.112	0.107	0.102	0.098	0.095	0.092	0.080
3	0.201	0.190	0.180	0.172	0.165	0.159	0.153	0.149
4	0.286	0.270	0.255	0.243	0.232	0.222	0.214	0.208
5	0.374	0.352	0.332	0.316	0.301	0.287	0.275	0.265
6	0.464	0.436	0.411	0.389	0.370	0.353	0.338	0.325
7	0.558	0.522	0.491	0.466	0.443	0.423	0.405	0.389
8	0.657	0.614	0.578	0.549	0.524	0.502	0.481	0.463
9	0.766	0.722	0.684	0.652	0.625	0.600	0.578	0.558
10	0.912	0.890	0.872	0.856	0.842	0.829	0.817	0.806

Chapter Two
TIMBER BRIDGES

SECTION I. GENERAL REMARKS

The type described in this Chapter is essentially the timber bridge of moderate span, designed for heavy modern traffic. The design methods and the examples refer mainly to the classical conservative style, but in addition, some information is also given on the latest attainments in this branch of engineering works; for instance, the so-called "split-rings" (*Tuchscherer System*), the plywood I-beams, and various advanced methods used in detailing the trusses in the buildings of the Golden Gate Exposition.

SECTION II. DESIGN OF TIMBER BRIDGES

1. Introductory

Timber was commonly employed in bridge construction in the past, when the cost of steel was prohibitive, whilst on the other hand the elastic theory of arches was unknown, and the design of masonry arches of wider spans was irrational and therefore, as often as not, uneconomical. Nowadays, timber is but seldom used in permanent works (except under emergency circumstances, i.e. during war), although it is still frequently embodied in temporary structures; such as falsework, centring, temporary means of communication, etc.

The *disadvantages* of timber bridges are:

(1) Their short life and expensive maintenance.

(2) The awkward, large sizes of the timber elements required to carry the heavy loads imposed by modern petrol-driven road traffic.

The *advantages* being:

(1) Short time required for erection.

(2) Simplicity of workmanship and adaptability to local conditions.

(3) Possibility of re-using the same timber members for several successive constructions, for temporary bridge spans.

2. Details of Design of Timber-work, Carrying Calculated Stress

The essential points in designing wooden bridges, and similar timber-work that may be subject to calculated stress, are:

(1) The resistance of wood is different in different directions.

(2) It depends upon the angle between the direction of the force applied (see Fig. 37) and the direction of the fibres (grain).

TIMBER BRIDGES

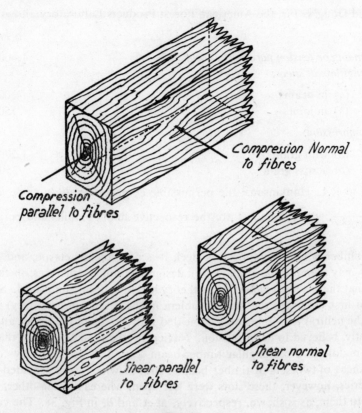

Fig. 37.

The following table gives a set of reasonable stress limits incorporating these principles.

Permissible Stresses in kg/cm²

	A Conservative continental practice (Germany)		B Conservative American (railway) practice	
	Oak	Pitch-pine	Oak	Pitch-pine
Tension and Bending	100–120	80–100	96.8	96.8
Compression in short blocks:				
(a) parallel to grain	80–100	60	86	73
(b) normal to grain	45	15	39.6	15
Shear:				
(a) parallel to grain	20–15	15–10	18.5	15
(b) normal to grain	80–90	60–70	—	—

Note 1. It is specified that the B limits may be raised 50 per cent for temporary structures.

Note 2. For columns, when the length, l, exceeds 15 times the width, d, the American limits are:

    For oak $115(1 - l/60d)$ kg/cm².

    For pitch-pine $96.8(1 - l/60d)$ kg/cm².

For selected Douglas Fir, the American Forest Products Laboratory allows:[1]

	lb/sq in
Bending or tension parallel to grain	1,600
Longitudinal shear:	
(*a*) in beams	100
(*b*) in joints	150
Compression:	
(*a*) with grain	1,200
(*b*) across grain	350

According to R.L. Hankinson,[2] the permissible compression at an angle θ to the grain is $\frac{PQ}{P \sin^2 \theta + Q \cos^2 \theta}$, where P and Q are the respective limits parallel and perpendicular to the grain.

Although timber is not as uniform as steel, it is nevertheless elastic, and the common bending theory may therefore be employed in designing timber structures; in fact, it was on timber beams that the practical value of this theory had first been established. Since the time of Galileo (who first discussed the beam problem in the seventeenth century) it was at first supposed that the neutral axis in bending coincided with the top of the beam; all the material was, consequently, believed to be in tension. Navier, in about 1820, proved that the neutral axis lay in the middle of a rectangular timber-beam section. This he did by comparing the ultimate resistances of two sets of timber beams, both of which were provided with vertical slots. In one series, however, these slots were left open, whereas in the other, wedges were firmly driven into them, as is shown, respectively, at *A* and *B,* in Fig. 38. The carrying capacity of the beams with wedges was found to be much greater; which proved the existence of compression in the upper fibres of the beam.

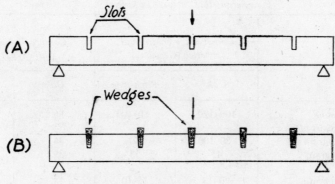

Fig. 38.

The factor of safety for wood is assumed to be from 5 to 6. This is greater than the coefficient adopted for steel, for the following reasons:

1. The stress distribution is more involved in timber-work.

[1] *Proceedings Am. Soc. C.E.,* Volume 73, No. 5, p. 585.
[2] Investigation on Crushing Strength, etc., Circular 256, U.S. Air Service Information, 1921.

TIMBER BRIDGES

2. Owing to differences in grain structure, timber is less homogeneous than steel, which creates an additional factor of uncertainty.

3. The time factor affects timber-work to a greater extent than steel, which means that the strength of timber structures gradually decreases, particularly if exposed to weather action.

4. Air humidity causes deformation of timber-work (swelling) which, in certain types of work, may eventually result in secondary stresses of some importance.

3. Application of Main Principles to the Design of Timber Structures

We will now explain how these main principles affect the design of the various elements of a timber-work.

Tension Ties. According to well-established conservative European practice, tensile stress must not be transmitted from one timber to another by bolts at right angles to the axes of the latter. The arrangement shown in Fig. 39 is therefore considered to be defective, though it may be frequently used in less-important work.

According to this theory, if bolts are employed in connections, calculated stress of practical importance must be transmitted parallel to their axes only (except in diagonal bracings of trestles, where the entire strength of the timber section is not utilized). With reference to Fig. 39, which is intended to serve as an illustration of the principle, the permissible force in the tie is:

$$20 \times 24 \times 80 = 38,400 \text{ kg}$$

whereas the carrying capacity of the bearing surface of three 3/4 inch bolts is only $3 \times 1.9 \times 24 \times 60 = 8,208$ kg, or about 20 per cent only of the total strength of the timbers. This shows that the design is basically unsatisfactory.

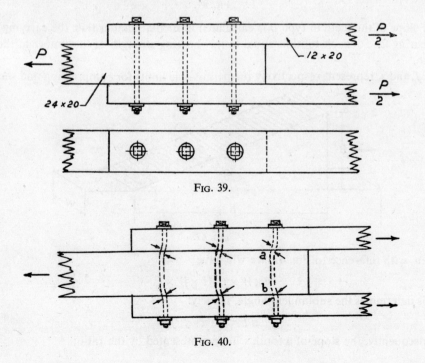

Fig. 39.

Fig. 40.

Practically, the maximum local bearing stress will be greater than the average calculated value, because the bolt must bend, and stress-concentrations will consequently occur at the points *a* and *b* (see Fig. 40); with the result that the steel will cut the timber as a "knife" at these spots. This will render the difference between the respective resistance of timber and bolt still more pronounced, making the design still less efficacious.

Instances of the connections designed according to sound theory are shown in Fig. 41.

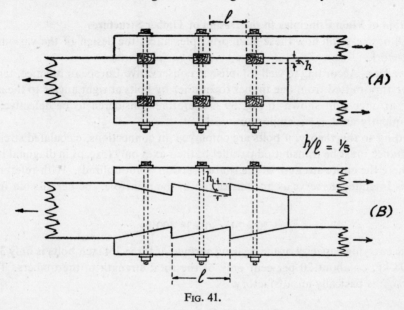

FIG. 41.

The slope of the teeth in type B is calculated in such a manner that the carrying capacity of the bearing face of each tooth is equal to the shearing strength of the timber at the base of this tooth.

Let f_c and f_s represent respectively the permissible limits for compression and shear.

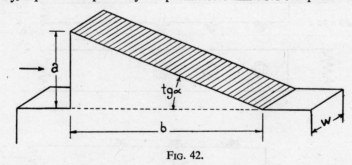

FIG. 42.

Then, with reference to Fig. 42, we will have

$$a \times W \times f_c = b \times W \times f_s.$$

This determines the subtangent angle, namely:

$$a/b = \tan \alpha = f_s/f_c.$$

Consequently, the slope of a tooth will be represented by the ratio:

TIMBER BRIDGES

$$\frac{\text{permissible stress for shear}}{\text{permissible stress for compression}}.$$

It follows that, for oak, this slope may vary from $\frac{15}{100}=0.15$ to $\frac{20}{80}=0.25$, whereas for pitch-pine it will be from $\frac{10}{60}=0.16$ to $\frac{15}{60}=0.25$.

A mean value of one-to-five is frequently used, as a general guide for practice. The same principle must obviously apply to the design of keys in Fig. 41A; that is to say, the width of the keys must be two-and-a-half times their half-height, and the spacing five times their half-height.

It follows that:

1. The net section of the ties is materially reduced owing to the use of teeth or keys at connections.

2. An extra length of timber is required, in order properly to develop the tensile force in the joints of timber ties.

It is therefore clear that the design of tension members in structural timber-work carrying calculated stress is rather expensive, and most often also the full strength of such members is not utilized completely.

This difficulty led to the idea of introducing light steel tensile elements into heavy timber framework. One of such devices, the well-known Howe truss, is still occasionally used at present. It is schematically illustrated in Fig. 43.

The timbers forming the lower chord are relatively long, and the percentage of overlapping is consequently less important than is the case with the verticals, which being much shorter, would come out very expensive, had they been made in wood. These verticals are therefore made of steel.

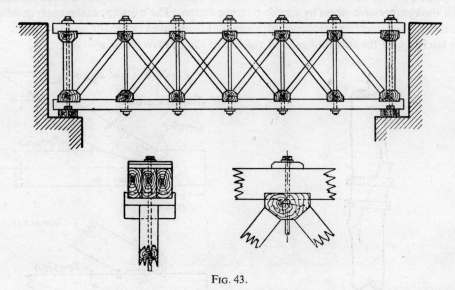

Fig. 43.

An additional advantage of the system lies in the possibility of adjusting the camber of the truss by tightening the nuts in the connections of the vertical ties.

Further examples of timber-trusses will be given later on, in discussing the results of the Golden Gate tests.

4. Types of Standard Connections

Conservative Continental Type. Typical designs for inclined and vertical connections are shown in Figs. 44 to 46. Here again, as in the preceding examples, the conservative continental practice is assumed to be followed.

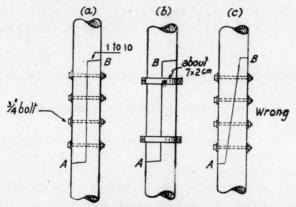

Fig. 44.—Vertical splices.

The characteristic simplicity of the compression splice, illustrated in Figs. 44 (*a*) and 44 (*b*), as opposed to the inherent complication of tension splices, which we have previously considered, is a distinct characteristic of the detailed design of important engineering timber framework. In fact, two compression members, abutting against each other, would have sufficed to transmit the compressive stress by simple bearing action. The overlap, under such conditions, is not intended for carrying calculated stresses, but is required either for additional resistance against buckling, or for ensuring general rigidity of the framework as a whole.

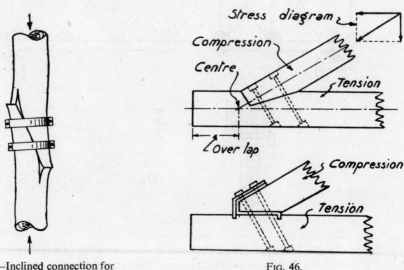

Fig. 45.—Inclined connection for compression members.

Fig. 46.

TIMBER BRIDGES

Over the length of the overlap (*AB* in the figure), the surface of contact must preferably be parallel to the main axis of the beam, as shown in Figs. 44 (*a*) and 44 (*b*), but not sloping as in Fig. 44 (*c*), for the latter arrangement, apart from reducing the working bearing area, creates a predisposition towards the formation of cracks, such, for instance, as that shown in Fig. 45.

Another point characteristic of rational timber-work design is that, although theoretically a compressive stress *P* (see Fig. 47) can be transmitted from *A* to *B* by direct contact as shown in sketch (*a*), yet in practice the efficiency of the transmission is much improved by the introduction of an intermediate bearing member; this may either be a harder class of timber, as shown in Fig. 47 (*b*), or (especially in America) a metal plate. The reason is explained in Fig. 47 (*c*); in fact, the natural micro-structure of timber is a succession of hard and weak fibres, and, unless a bearing block, such as *m* in Fig. 47 (*b*), is introduced between the abutting surfaces of the two timbers, the harder elements of one piece tend to penetrate into the less-resistant layers of the other timber; thus causing a general dislocation of the stratified micro-structure, which in turn results in the sagging of the whole structure.

In inspecting the sloping connections in Fig. 46, attention is called to the fact that the compression member does not need to project very far beyond the centre of the connection; which means that the design of this member is very much the same as if it had been a part of an ordinary compression splice; on the other hand, a timber element in tension must be provided with a considerable overlap, since the tensile stress could not be effectively developed otherwise. Thus, the characteristic difference between timber details in compression and in tension remains also valid for sloping connections.

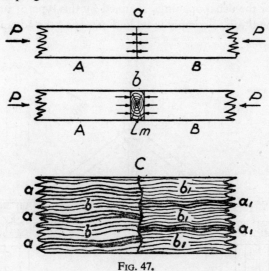

Fig. 47.

The two structural devices shown in Figs. 48 and 49 were frequently used in the nineteenth century. They were indeed eminently suitable for the type of traffic characteristic of the period—weighing up to two or three tons per vehicle—but for the heavier modern traffic they are definitely lacking in rigidity.

Take for instance the type shown in Fig. 48. A load applied at *A* causes the bearings *B* and *C* to be shifted outwards as shown by arrows. So long as the overall rigidity of the timber-

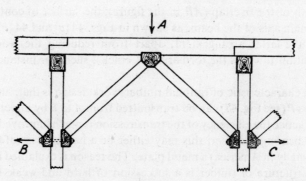

Fig. 48.

work suffices to withstand this tendency, well and good; but, after a certain limit is exceeded, the joints B and C begin to yield, and the main girder is then at once crushed at A.

As a temporary measure one could tie the beams B and C together by a steel bar or by substantial wires. This indeed is often done under emergency circumstances, to allow a heavy lorry or road roller to cross such a bridge. Bars or wires of this type have, however, a tendency to sag and cannot be depended upon to work as a permanent feature of the design. On the other hand, the trestle bridge in Fig. 49 lacks rigidity in a direction at right angles to the plane of the drawing. Girders of this type must therefore be counterbraced right across the roadway, and in order to do this they have to be made sufficiently high to let the rolling traffic pass below the bracings. For the usual openings spanned by this type of bridge, this requirement may be an awkward condition to comply with, and can spoil the advantages of the system altogether.

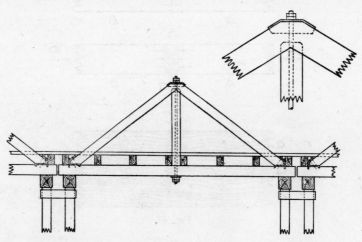

Fig. 49.

Broadly speaking, the simple but robust type of ordinary timber girder, provided it is properly calculated and rationally designed as explained in the following subsection, is better suited to the conditions imposed by modern heavy loading than the more involved "fancy" systems of complicated timber framework popular with the designer of the last century.

TIMBER BRIDGES

5. Trestles and Supports

Timber spans may rest either on masonry piers and abutments, or on wooden trestles and piles. Application of the latter principle for permanent work is not recommended in canals, because timber decays at a very rapid rate if subjected alternately to the effect of water and air. This alternation necessarily affects timber piles in irrigation canals, in which the water-level must vary very often. Nevertheless, there are still many structures of this type on the

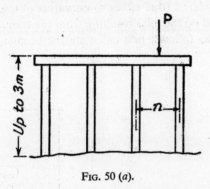

FIG. 50 (a).

canals of existing systems, although nowadays timber bridges are being gradually replaced by reinforced concrete designs. The sketch in Fig. 50 (a) gives the simplest type of a timber trestle, suitable for a height up to three metres above ground-level (or canal-bed); more involved and higher trestles are illustrated in Figs. 50 (b) and (c) to 52. In the latter two cases, i.e. Figs. 51 and 52 the trestles must also be counterbraced longitudinally, i.e. at right angles to the plane of the drawing.

Calculation of Trestles. The precise theoretical value of the permissible vertical loading on a trestle is a function of the relative elasticities of the cap and piles, but for practical purposes the cap may be assumed to be sectioned and freely supported at each pile, the effect of continuity of the cap being thus disregarded for sake of simplicity. The stresses in the cap and

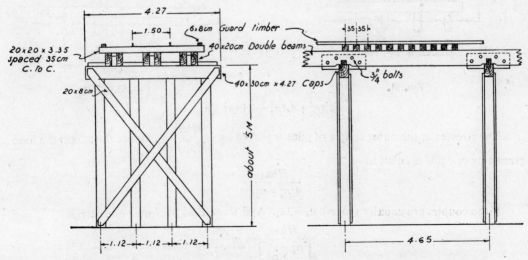

FIG. 50 (b) and (c).—Standard type of wooden trestle bridge of the New York R.W.

the loading of the piles will then be easily ascertained; namely, the bending moment for the cap (see Fig. 50 (a)) will be equal to $\frac{(n-a)\,aP}{n}$ and the load on the pile $\frac{Pa}{n}$ in which a is the distance between load and pile and n the span. For inclined piles multiply the latter formula by $\frac{1}{\cos \alpha}$, where α is the angle between the vertical and the pile. In the cases in which there is a horizontal thrust to be considered (due either to curvature of track or to wind action) there will be an additional overload on the piles resulting from the transverse bending moment. This will be taken into account as follows: first calculate the moment of inertia of the piles at surface-level (see Fig. 52):

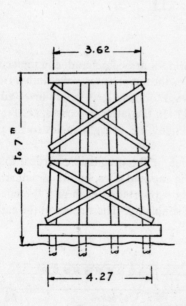

Fig. 51.

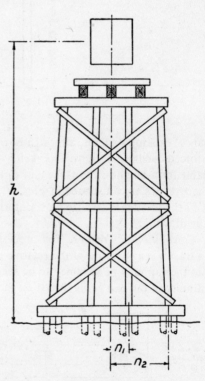

Fig. 52.

$$I = 4An_1^2 + 4An_2^2 = 4A\,(n_1^2 + n_2^2).$$

The stresses in the outer couple of piles will then be $\frac{Hhn^2}{2A[n_1^2 + n_2^2]}$. Hence the maximum load carried by one pile is equal to

$$\frac{Hhn_2}{4(n_1^2 + n_2^2)}.$$

If the couples are equally spaced: $n_2 = 3n_1$. And therefore, the outer pile carries:

$$\frac{Hhn_2}{4\left[\frac{n_2^2}{9} + n_2^2\right]} = \frac{9}{40}\frac{Hh}{n_2}.$$

TIMBER BRIDGES

SECTION III. EXAMPLE OF DESIGN OF A TIMBER SPAN

1. Flooring, Footpath, etc.

It is required to design a timber span for the standard 20-ton lorry;[1] clear span 7 metres; roadway 5 metres wide and two footpaths 1.0 metre each.

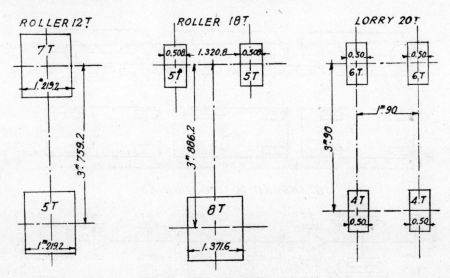

Fig. 53.—Standard live loads.

Calculation of Flooring. The timber flooring constitutes both the wearing surface and the carrying element of the deck, and may therefore be either simple or double. In the first case, the same material performs both duties, whereas in a double flooring the upper part is a "cuirasse" only, which protects the lower part from damage, but does not participate in carrying a calculated load. If a simple flooring is used, its thickness should be increased by 2 to 3 cm above the theoretically calculated dimension, in order to provide a margin for wear.

Usual Size. This is 5 to 8 cm (2 to 3 inches) for the upper flooring, and 10 to 15 by 20 to 30 cm (4 to 5 inches by 8 to 12 inches) for the elements of the lower flooring.

Cross-Girders. No cross-girders are used in small timber bridges.

Span of Flooring. Consequently, the span of the flooring is equal to the spacing of the main girders.

The lower or main flooring is usually placed at right angles to the main girders. The upper flooring may then be either longitudinal or diagonal.

The standard thickness of the main flooring being assumed to be known, the spacing of the girders becomes a function thereof.

For our example, suppose the upper flooring to be 6 cm and the lower flooring 15 × 25 cm.

The following notation will be adopted (see Fig. 56);

x = span of flooring in centimetres (spacing of girders).

n = width of wheel in centimetres.

---

[1] For facility of reference a selection of types of modern live loads is shown in Figs. 53 to 55. These are the most common types, although heavier loads are sometimes specified.

P = weight on wheel.

Bending Moment. Assume the flooring to be simply supported on the beams; according to Fig. 56 we will then have for the live moment:

$$M_e = \frac{P}{2}\left(\frac{x}{2} - \frac{n}{4}\right).$$

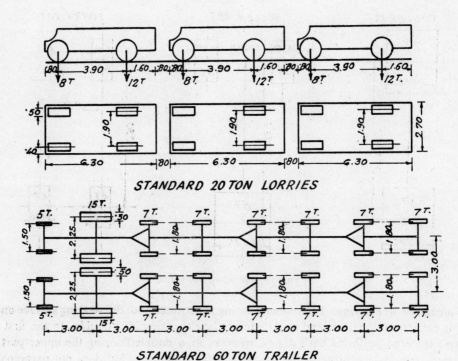

FIG. 54.

For the standard, 20-ton lorry (see Fig. 53), $P = 6{,}000$ kg and $n = 50$ cm. It follows that the live moment, M_e, is equal to

$$M_e = \frac{6{,}000}{2}\left(\frac{x}{2} - \frac{50}{4}\right) = 3{,}000\left[\frac{x}{2} - 12.5\right]$$

$$= 1{,}500x - 37{,}500 \text{ kg} \times \text{cm}.$$

Structural Weight. The weight of a timber flooring is relatively very small (M_d about one per cent of M_e). It may therefore be disregarded in the preliminary calculation of stresses.

Section Modulus. In the case of a double flooring, the load of the wheel is assumed to be transmitted to two planks of the lower flooring, and therefore the section modulus is equal to

$$Z = \frac{2 \times 25 \times 15^2}{6} = 1{,}875 \text{ cm}^3.$$

The permissible stress for bending, f, is assumed as 80 kg/cm². Consequently:

$$1{,}500x - 37{,}500 = 1{,}875 \times 80.$$

TIMBER BRIDGES

Solving

$$x = \frac{1{,}875 \times 80 + 3{,}750}{1{,}500} = 125 \text{ cm.}$$

This fixes the spacing of the main girders.

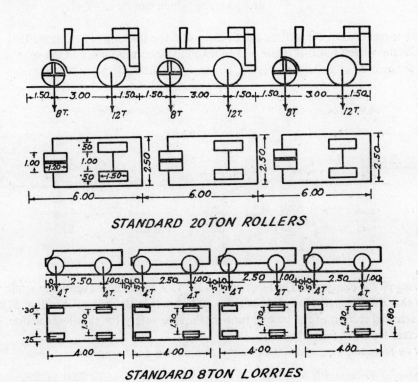

Fig. 55.

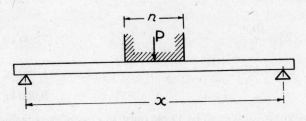

Fig. 56.

Verification of Stresses. We now include the structural weight, p_d:

$$M_e = \frac{P}{2}\left(\frac{l}{2} - \frac{n}{4}\right) = \frac{6{,}000}{2}\left(\frac{125}{2} - \frac{50}{4}\right) = 150{,}000 \text{ kg} \times \text{cm}$$

$$p_d = 50 \times (15 + 6) \times 0.0008 = 0.84 \text{ kg/cm}$$

$$M_d = \frac{0.84 \times 125^2}{8} = 1{,}640 \text{ kg} \times \text{cm}$$

$$M_{total} = M_e + M_d = 150{,}000 + 1{,}640 = 151{,}640 \sim 152{,}000 \text{ kg} \times \text{cm}$$

$$f = \frac{152{,}000}{1{,}875} = 81.05 \text{ kg/cm}^2 \text{ which can be allowed.}$$

Another method for calculating the thickness of the flooring is to assume that one plank only carries the wheel; but this plank is then taken to be continuous, and the span is measured not between centres of girders, as before, but between their edges (clear span).

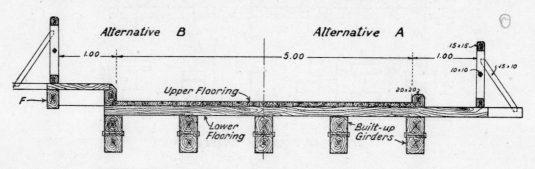

Fig. 57.

Flooring on the Footpaths. The length of planks 15 × 25 cm may eventually attain 9 metres; it follows that the footpaths may be built as cantilevers (see alternative A in Fig. 57).

The selected section of the boards furnishes ample safety for cantilever action. This will be made clear from the following:

Load on footpath, assumed 600 kg/sq met, which is indeed ample.

Cantilever.	Total width of bridge	7.00 metres
	Width supported by main girders 4 × 1.25	5.00 metres
	Difference	2.00 metres
	Length of cantilever	1.00 metre

Consequently:

$$M_e = \frac{p_e l^2}{2} = 0.06 \times 25 \times \frac{100^2}{2} \qquad 7{,}500 \text{ kg} \times \text{cm}$$

$$M_d = \frac{p_d l^2}{2} = (25 \times 15 \times 0.0008) \frac{100^2}{2} \qquad 1{,}500 \text{ kg} \times \text{cm}$$

$$\text{Total} \qquad 9{,}000 \text{ kg} \times \text{cm}$$

The effective thickness is taken to be 12 cm only, in view of possible wear in future. Hence:

$$Z = \frac{25 \times 12^2}{6} = 600 \text{ cm}^2 \text{ and } f = \frac{9{,}000}{600} = 15 \text{ kg/cm}^2.$$

In the case if the length of the available timber boards does not allow this solution to be adopted, alternative B in Fig. 57 may be chosen. An additional girder F is then used to support the outer edge of the footpath.

TIMBER BRIDGES

2. Main Girders

The wheel is supposed to be located directly over the beam, which therefore carries its total weight.

Live Moment (*see Fig. 58*). To find the maximum live-load moment, place the centroid of the loading diagram and the main load at equal distances from the middle of the span. The following formulae are developed for the 20-ton standard lorry according to this criterion. The maximum live-load moment is then equal to $M_e = 3.0 \times A$ kg × metres, where $A = \dfrac{10,000 \times 3.0}{7.60} = 3,950$ kg is the reaction at the right-hand bearing.

Hence:

$$M_e = 3.0 \times 3,950 = 11,850 \text{ kg} \times \text{metres}.$$

Note. Instead of the exact solution based on the above criterion, as shown in Fig. 58, we might have placed the six-ton wheel in the centre of the span, as shown in Fig. 58 (*a*); then

$$M_e = 3,000 \times 3.8 = 11,400 \text{ kg} \times \text{metres}.$$

This means a $100 \dfrac{450}{11,850} = 4$ per cent difference only. Thus, for a preliminary, rapid estimate, the approximate solution is sufficiently accurate.

Dead Moment due to Structural Weight:

Structural weight per metre run—Flooring—$0.21 \times 1.25 \times 800 = 210$ kg
Assumed weight of beam $\quad\;\; 200\ ,,$
Total $\qquad\qquad\qquad\qquad\quad \overline{410\text{ kg}}$

or roughly, 400 kg/metre.

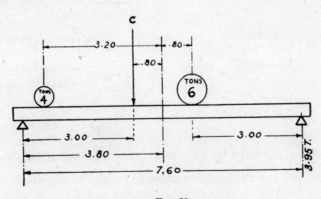

Fig. 58.

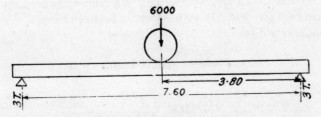

Fig. 58 (*a*).

Hence:
$$M_d = 400\,\frac{7.6^2}{8} = 2{,}890 \text{ kg} \times \text{metres}.$$

$$M_{total} = 11{,}850 + 2{,}890 = 14{,}740 \text{ kg} \times \text{metres} = 1{,}474{,}000 \text{ kg} \times \text{cm}.$$

The largest size of beam usually available—under normal market conditions—is about 35 × 35 cm (14 × 14 inches). Accordingly, the maximum section modulus for one beam is:

$$Z = \frac{35^3}{6} = 7{,}150 \text{ cm}^3.$$

The corresponding stress is: $f = \dfrac{1{,}474{,}000}{7{,}150} = 206 \text{ kg/cm}^2$, which materially exceeds the permissible limit. Consequently, a single beam will not suffice in our case.

Should two beams be placed one upon the other, the resistance will be doubled; thus, the stress will be:

$$f = 103 \text{ kg/cm}^2, \text{ which is, again, too high.}$$

The full-section modulus for the combined section of the two beams can be included in the calculation only in the case, if the sliding of the upper beam upon the lower one is prevented. To explain this contention, attention is called to Figs. 59 and 60.

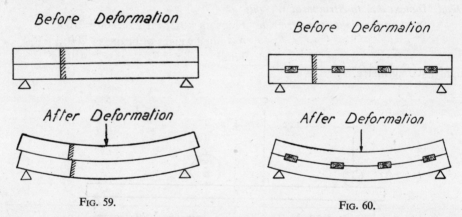

Fig. 59.   Fig. 60.

The formula $f = \dfrac{M}{Z}$ is derived from the assumption that the normal sections of the beam remain true planes after the elastic deformation has taken place, which is not the case in Fig. 59, but is realized in Fig. 60, owing to the presence and action of keys. The two beams combined in that manner form then a "built-up" girder.

Calculation of the Built-up Girder. We will assume a built-up girder consisting of two timber beams 30 × 30 cm each (see Fig. 61), with N keys in between them.

The section modulus will then be:

$$Z = \frac{60^2\,30}{6} = 18{,}000 \text{ cm}^3.$$

and the stress

$$f = \frac{1{,}474{,}000}{18{,}000} = 82 \text{ kg/cm}^2,$$

TIMBER BRIDGES

which can be allowed. The number and position of the keys must now be calculated from the diagram of the shearing force.

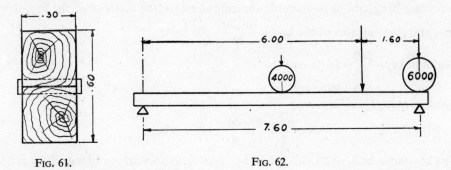

Fig. 61.   Fig. 62.

Calculation of Shear:

Maximum live shear (see Fig. 62) at support $= A_e = \dfrac{10,000 \times 6.0}{7.6} = 7,900$ kg

„ dead „ „ „ „ „ $= A_d = 400 \times \dfrac{7.6}{2} = 1,520$ „

Total, A $= 9,420$ kg.

Shearing Stress $S_1 = AF/Ib$. For a rectangular section, this formula yields one-and-a-half times the average shearing stress:

$$(f_s)_1 = 1.5 \frac{A}{w} = 1.5 \frac{9,420}{1,800} = 7.85 \text{ kg/cm}^2,$$

where w = area of section of built-up beam, $60 \times 30 = 1,800$ cm².

Shear in the middle of the span:

Live shear $= \dfrac{6,000}{2} = 3,000$ kg

Dead shear $=$ 0 „

Total $A_2 = 3,000$ kg

Shearing stress $(f_s)_2 = 1.5 \dfrac{3,000}{1,800} = 2.5$ kg/cm².

Shear per centimetre length of beam $b \times f_s$:

At support $a_0 = b \times (f_s)_1 = 30 \times 7.85 = 235.5$ kg/cm
At middle $a_1 = b \times (f_s)_2 = 30 \times 2.50 = 75.0$ „ „

Total longitudinal shear for half the span:

$$S_T = \frac{b\,[(f_s)_1 + (f_s)_2]}{2} \times \frac{l}{2} = \frac{235.5 + 75}{2} \times 380 = 59,000 \text{ kg}.$$

Keys. The keys are subject to compression at right angles to the grain and are therefore made of oak or similar hard timber, capable of resisting a substantial compressive stress across the grain.

Assuming 15 cm depth and 45 kg/cm² permissible compressive stress across grain, the bearing resistance of a key is

$$R = \frac{15}{2} \times 30 \times 45 = 10{,}120 \text{ kg.}$$

Adopting 20 kg/cm² as permissible shearing stress for the material of the key, the required horizontal shearing area of the key is $\frac{10{,}120}{20} = 506$ cm².

Width of key $= \frac{500}{30} = 16.25$ cm.

        Adopted section 15×17 cm

Number of keys per half-span:

$$N = \frac{S_T}{R} = \frac{59{,}000}{10{,}120} = 5.8 \text{ or 6 keys.}$$

The keys must be boiled in oil before being used, to prevent shrinking. They are slightly wedge-shaped; the slope being 1 to 3 per cent to ensure tight fitting, when driven in position by a wooden mallet.

3. Spacing of Keys

The keys are spaced depending on the shear diagram, so as not to exceed the shearing limit in the timber; i.e. they must be located in such a manner that each key should take an equal part of the total shear, S_T.

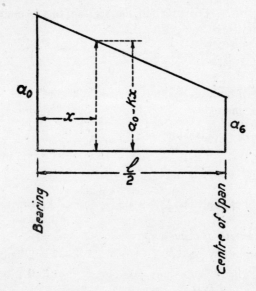

Fig. 63.

The shear diagram is shown in Fig. 63, the ordinates represent in this chart the shearing stress in kilograms per centimetre length of the beam, and the abscissae are the lengths in centimetres.

It follows that the area of the diagram gives the total shear, S_T, in kilograms. In order to space the keys correctly, this area must now be divided into equal sections, and the keys must fall in line with the centroids of these sections.

This is done as follows:

TIMBER BRIDGES

Let the slope of the diagram be $K = 2(a_0 - a_b)/l$.

The area of the shearing stress diagram corresponding to a given value of x (distance to the centre of the end bearing) is

$$\frac{a_0 + (a_0 - x \tan \alpha)}{2} \times x = a_0 x - \frac{x^2}{2} K.$$

To divide the area of the diagram into N parts, the distance to the nth division must correspond to an area nS_T/N. We will therefore have

$$\frac{nS_T}{N} = a_0 x_n - \frac{x^2}{2} K \quad \text{and} \quad \frac{x^2}{2} K - a_0 x + \frac{nS_T}{N} = 0.$$

It follows that

$$x^2 - \frac{2a_0}{K} + \frac{2nS_T}{KN} = 0.$$

Solving

$$x = \frac{a_0}{K} - \sqrt{\left(\frac{a_0}{K}\right)^2 - \frac{2nS_T}{KN}}.$$

In our case we have:

$$a_0 = 233.5 \text{ kg/cm}$$
$$a_b = 75.0 \text{ ,, ,,}$$
$$l = 760.0 \text{ cm}$$
$$S_T = 59{,}000.0 \text{ kg}$$
$$N = 6.0$$

Applying the formula we find:

n	x	$x_n - x_{n-1}$	n	x	$x_n - x_{n-1}$
1	44 cm	44 cm	4	206 cm	61 cm
2	92 ,,	48 ,,	5	281 ,,	75 ,,
3	145 ,,	53 ,,	6	380 ,,	99 ,,

It remains to find the centroids of the divisions.

This operation includes two stages: (*a*) calculate the ordinates corresponding to the limits of the divisions, and (*b*) find the centroids accordingly.

The slope of the diagram is:

$$K = 2 \frac{a_0 - a_b}{l} = \frac{233.5 - 75.0}{380} = 0.417 \text{ kg/cm.}$$

For the first operation we may use the following equation $a_n = a_0 - K x_n$. This will give:

$$x_1 = 44 \qquad a_1 = 215.2 \text{ kg/cm}$$
$$x_2 = 92 \qquad a_2 = 195.1 \text{ ,, ,,}$$
$$x_3 = 145 \qquad a_3 = 173.0 \text{ ,, ,,}$$
$$x_4 = 206 \qquad a_4 = 147.6 \text{ ,, ,,}$$
$$x_5 = 281 \qquad a_5 = 116.3 \text{ ,, ,,}$$
$$x_6 = 380 \qquad a_6 = 75.0 \text{ ,, ,,}$$

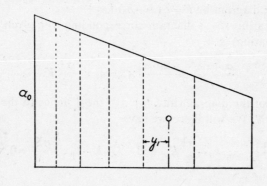

Fig. 64.

The formula for the second operation (centroid of trapezium) will be (see Fig. 64):

$$y = \frac{a_{n-1} + 2a_n}{a_{n-1} + a_n} \times \frac{x_n - x_{n-1}}{3}.$$

The calculation carried out according to this equation is shown in the table below.

Section	$a_{n-1}+2a_n$	$a_{n-1}+a_n$	$\dfrac{a_{n-1}+2a_n}{a_{n-1}+a_n}$	$\dfrac{x_n-x_{n-1}}{3}$	y	$y+x_{n-1}$
1	663.9	448.7	1.26	14.7	18.5	18.5
2	605.6	410.3	1.48	16.0	23.7	67.7
3	541.1	368.1	1.48	17.7	26.2	118.2
4	468.2	320.6	1.46	20.3	29.6	174.6
5	380.2	263.9	1.48	25.0	37.0	243.0
6	266.3	191.3	1.39	33.0	45.9	326.9

The same result might have been obtained graphically, as shown in Fig. 65. The method adopted is a well-known computation, calling for no comment.

Checking the Shearing Stress. The smallest distance between the keys is found from the last column of the table to be $67.7 - 18.5 \sim 49$ cm. From this distance, 17 cm are removed (see Fig. 66) to accommodate the key; thus, the shearing stress is carried by a length of 32 cm and is therefore equal to

$$\frac{10,120}{30 \times 32} = 10.6 \text{ kg/cm}^2.$$

This is less than 15 kg/cm² which is believed to be permissible for the material of the beam working in shear parallel to grain, and is therefore acceptable.

This is the final check. Had the permissible limit for shear been exceeded, the entire computation might have to be repeated all over again.

End Bearings. In order properly to distribute the concentrated reactions over the masonry supports, a cushion, consisting of two transverse beams, each 20×20 cm in section, is placed on the abutment, as shown in the drawing (see Fig. 67).

TIMBER BRIDGES

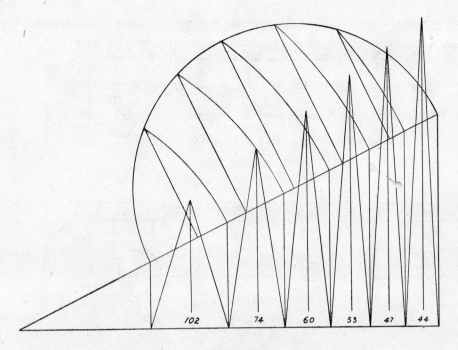

Fig. 65.

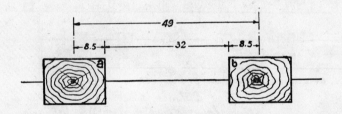

Fig. 66.

Thus, the bearing stress prependicular to the grain is $\dfrac{9{,}420}{30 \times 40} = 7.9$ kg/cm².

The calculation of the built-up girder is thus completed. It must be noted, however, that some designers prefer to use teeth instead of keys. The calculation will then be substantially the same but in order to develop an equal resistance slightly deeper timbers must be used; or alternatively, if the same sections are employed, the carrying capacity of the beam will be smaller. A built-up girder with teeth instead of keys, but embodying in all other respects the results of the stress analysis—presented in the foregoing pages, is shown in Fig. 67 as alternative (*a*).

From the practical standpoint it is rather important to realize that whatever alternative (i.e. keys or teeth) is used, special measures must be taken to prevent the beam from sagging; it must therefore be pre-stressed and provided with the required camber.

In this connection it may be of interest to mention that the routine of pre-stressing was common among timber craftsmen on the Continent (particularly in Galicia, on the wooded

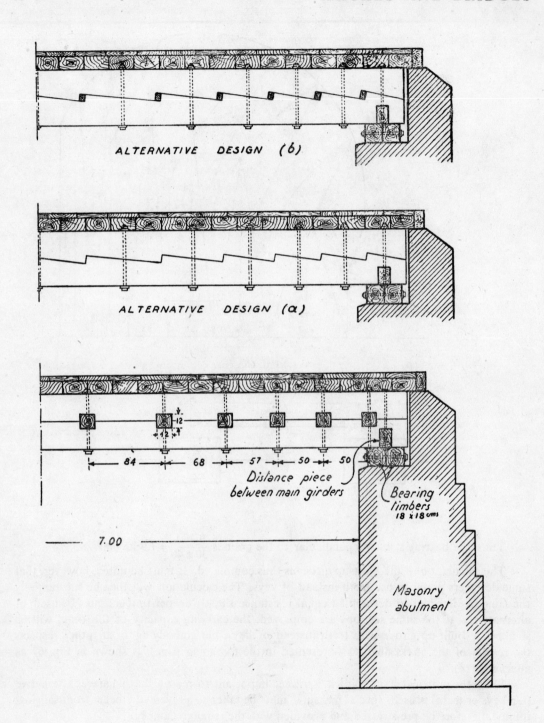

Fig. 67.

TIMBER BRIDGES

slopes of the Carpathian Mountains) much earlier than it had ever been thought of by advanced designers, in connection with reinforced concrete designs.

The process is as follows:

(*a*) The two beams are erected, one on top of the other, without keys or teeth and are loaded to their full (or almost full) carrying capacity, i.e. double the capacity of a single beam;

(*b*) The position of the keys or teeth is then marked out on the surface of the timber;

(*c*) The timbers are taken down and finished in accordance with these markings;

(*d*) They are then re-erected and re-loaded, and completely assembled, fixing the keys in the corresponding grooves, or alternatively, adjusting the teeth in their final position, and bolting the beams together;

(*e*) The load is removed and the beams are turned upside-down, so that the built-up girder should be slightly convex on its upper surface.

The objective and the underlying reasons of this programme are obvious, but there is no doubt that it is rather involved and requires experienced craftsmen, trained in this kind of work, to carry it out. It is also open to objection on the score that the camber of the bridge cannot be adjusted (increased?) after erection.

To avoid these objections a third solution is possible. It is represented as alternative (*b*) in Fig. 67, and supplies, in a way, a combination of teeth with keys. The latter, however, perform in this case the function of wedges, projecting substantially beyond the lateral faces of the main timbers. The built-up girder is erected in the usual way and the cambering and prestressing are then effected by driving home the wedges. These are made long enough to allow additional cambering six months or a year after erection.

This latter system has also the advantage of increasing the bearing capacity of the teeth (see p. 54 and Fig. 47).

SECTION IV. STRENGTHENING AN EXISTING TIMBER BRIDGE

This is rather a frequent problem encountered in modern maintenance practice. Suppose it were considered that the beams of an existing bridge were insufficiently strong to carry the new, heavy road traffic.

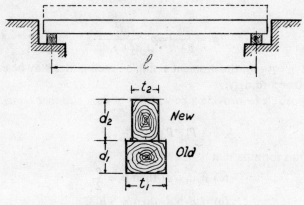

Fig. 68.

The simplest solution would then be to place, over the existing beam, an additional girder as shown by dotted lines in Fig. 68. To develop properly their combined strength, we should have used keys or teeth to join the old and new beams together, in one unit; but, in usual circumstances, this may be found too elaborate an operation to carry out in the field as an item of maintenance works. We will, consequently, consider the case when teeth or keys are out of the question, for practical reasons.

Suppose, then, the existing girder has a depth d_1 and its width is t_1. The corresponding values for the new beam are d_2 and t_2. It is then required to estimate the stresses in the combined structure, which will take place under a continuous load p, in case there are no keys or teeth. The crucial point is to find the respective parts p_1 and p_2 of the total load, taken respectively by the old and new beams. This problem cannot be solved from elementary, rigid statics, but requires application of the principle of elastic deformations; in fact, the deflections of the two beams must be the same in the middle of the span; consequently, we have the following two equations:

$$\frac{5 p_1 l^4}{384\ EI_1} = \frac{5 p_2 l^4}{384\ EI_2} \tag{1}$$

and

$$p_1 + p_2 = p. \tag{2}$$

Solving

$$p_1/I_1 = p_2/I_2$$

in which

$$I_1 = \frac{t_1 d_1^3}{12} \quad \text{and} \quad I_2 = \frac{t_2 d_2^3}{12}.$$

From this

$$\frac{p_1}{t_1 d_1^3} = \frac{p_2}{t_2 d_2^3} \quad \text{or} \quad p_1 = p_2 \frac{t_1 d_1^3}{t_2 d_2^3} = p_2 k$$

where

$$k = \frac{t_1 d_1^3}{t_2 d_2^3}.$$

It follows that

$$p_1 = (p - p_1)\, k = pk - p_1 k.$$

Hence

$$p_1 = \frac{pk}{1+k} = p \frac{t_1 d_1^3}{t_2 d_2^3 + t_1 d_1^3}.$$

Knowing p_1 and $p_2 = p - p_1$, the moment acting on each girder may be easily estimated and the stresses calculated accordingly.

It is frequently possible to provide a new girder of such size that $t_1 = t_2$. Then

$$p_1 = p \frac{d_1^3}{d_2^3 + d_1^3}.$$

It follows from this formula that

(a) if $d_1 = d_2$ then $p_1 = \dfrac{p}{2}$

(b) if $d_1 > d_2$ then $p_1 > p_2$

(c) if $d_1 < d_2$ then $p_1 < p_2$.

TIMBER BRIDGES

Thus the strongest of the two girders carries the greatest load.

Another method of strengthening an existing timber bridge is to use a steel bar forming, in combination with the old beam, a tension chord of a lattice girder (see Fig. 69). The original timber is then subject to compression only (apart from local bending).

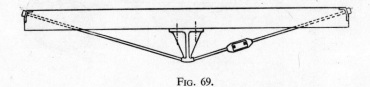

FIG. 69.

SECTION V. NEW TENDENCIES IN STRUCTURAL TIMBER DESIGN

1. Introductory

The traditional timber-design methods described in the foregoing pages have been universally applied in the earlier decades of this century. In addition to bridges designed for normal traffic, these methods have also been widely used in connection with hundreds (if not thousands) of major emergency bridges, built for military purposes during the two world wars. Economy, at that time, was possibly a secondary consideration as compared with the speed of construction, and efficiency. While these methods still constitute the classical solution of traditional timber-design problems, non-orthodox ideas have begun latterly to develop, as the result of a natural tendency towards reducing to a minimum the cost of labour and material.

2. "Connectors"

These new solutions first originated in Germany, but were later taken up and widely developed in the United States.[1]

The objective of the new method was to use the simple connection splice of Fig. 39, but to alter it in such a way as to satisfy the strict requirements of the rigorous stress theory. With this aim in view, a "connector", e.g. a "split ring", or a "shear plate", was introduced into the joint, rendering it thereby capable of transmitting a calculated stress, from one timber member to the other. These connectors are usually metal devices of such a shape that by using a simple tool the holes required for inserting them can easily be drilled in the mating surfaces of the corresponding timbers.

Broadly speaking, these devices are somewhat similar in principle to the keys shown in Fig. 41 A, with the difference, however, that instead of hard wood, they are made of steel or cast iron, and that they are circular in shape, in order to permit the use of drills for making the holes in which they are inserted.

The original design took the form of a flat iron bar bent to the required circular shape, as shown in Fig. 70. It constituted, therefore, a "split ring", and was known in Germany by the name of its inventor, the "Tuchscherer System".[2]

---

[1] An epoch-making American publication on the subject was: "Modern Connectors for Timber Construction" produced by the Forest Products Laboratory of the United States Department of Agriculture, in 1933 (U.S. Govt. Print. Office). Long-span timber beams of this type were mass produced during World War II and are believed to have contributed to the American war effort.

[2] Lewe, "The Calculation of the Split Ring of the Tuchscherer System", *Der Holzbau*, 1920, No. 20.

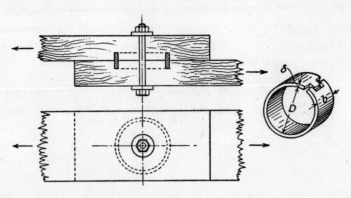

Fig. 70.

The purpose of splitting the ring was to keep the design effective, even after the shrinking of the material.

According to the usual German standards, the ratio of the diameter to the thickness of the metal, D/δ, is about 25. On the other hand, the width, b, of the plate is about five times δ.

With these proportions the maximum bending stress in the metal of the ring is $\sigma = 12.75\,p$, where p is the bearing stress between ring and timber; also, the shearing stress, τ, is 7.36 times p.

Assuming, then, $p = 60$ kg per sq cm (which is supposed to correspond to a safety factor of 4), the permissible load for a ring can be calculated from $P = Dbp$.

The main numerical data relating to this type of "connector" are summarized in the following table:[1]

D, mm	b, mm	δ, mm	Diameter of bolt, in	P, kg	D, mm	b, mm	δ, mm	Diameter of bolt, in
80	16	3.5	$\tfrac{1}{2}$	768	200	40	8	1
100	20	4	$\tfrac{5}{8}$	1,200	220	45	8	1
120	26	5	$\tfrac{5}{8}$	1,876	240	50	10	$1\tfrac{1}{8}$
170	29	6.5	$\tfrac{3}{4}$	2,436	260	52	10	$1\tfrac{1}{8}$
160	32	6.5	$\tfrac{3}{4}$	3,072	280	55	12	$1\tfrac{1}{4}$
180	36	8	$\tfrac{7}{8}$	3,888	300	60	12	$1\tfrac{1}{4}$

3. Golden Gate International Exposition in San Francisco

A considerable amount of experimental research work on the subject has been done in America, using the timber trusses taken down from the buildings of the Golden Gate International Exposition at San Francisco.[2]

Attention will first be called to Fig. 71, which shows one of the tested trusses, namely, that of the Alameda-Contra Costa Counties Building.

[1] Prof. Melan, *Der Brückenbau* (Vienna, 1922), Vol. I, p. 127.
[2] See report on "Tests of Timber Structures from Golden Gate International Exposition" published by the Committee of the San Francisco Section, A.S.C.E., on "Timber Test Program" in *Proc. Am. Soc. C.E.*, May 1947, pp. 573 and foll., from which most of the following information as well as Figs. 71–73 are taken.

TIMBER BRIDGES

It will easily be seen that this is but the Howe truss, to which reference has already been made on page 52 (see Fig. 43).

When tested to destruction, failure occurred "very suddenly in shearing off the table of the butt block at joint $L7$". The calculated shearing stress on the critical plane at the moment of failure was 474 lb per sq in, neglecting friction.

Failure at this relatively low stress might have been due to the abrupt change in section, at the upper of the re-entrant angles in the butt block, resulting in local concentration of high bending and shearing stress at the critical point.

Another truss, that of the Chinese building, is shown in Fig. 72. All truss members are here single rods, 2 inches in section. The compression diagonals are fixed on one side of the chords, while the tension verticals are on the opposite side. The lack of symmetry consequent on this detail of the design was the main cause of the failure of the truss. In fact, the eccentric connection of the truss members twisted the chords and produced in the members themselves bending stresses, resulting in rupture of material. At the moment of failure, the calculated breaking stress in the vertical $U1-L2$ was 1,030 lb per sq in in tension and 8,040 lb per sq in in compression.

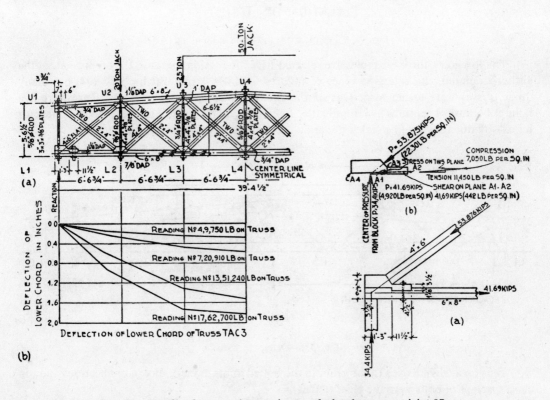

Fig. 71.—Details of truss and approximate calculated stresses at joint $L7$.

The third truss, illustrated in Fig. 73, failed, when after 1 hour and 13 minutes of testing with constant load, the joint $L5$ cracked audibly and longitudinal fissures 1/4 inch wide appeared in the chords, starting in the bolt-holes. These cracks were obviously due to local tension perpendicular to the grain, at the points where stress was transmitted to the material of the chord, by the bolts.

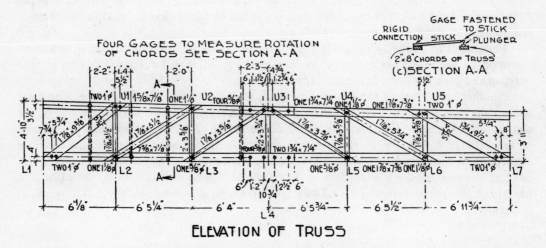

FIG. 72.—Typical truss, the Chinese building.

The foregoing three examples are selected here for citation, because they represent, in the author's opinion, characteristic types of failures found in structural timber-work; but many other trusses were tested also. Out of the twenty tests, on sixteen trusses, there were five failures because of ruptures in joint details, six failures at bolted joints and nine at connector joints—all due to types of stresses not anticipated in the traditional design methods.

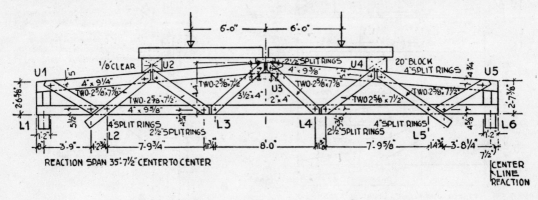

FIG. 73.—Details of truss.

Tension at right angles to the grain in the bolted joints, eccentricity of connections and too close spacings of bolts were the chief faults.

Still, the factors of safety derived from these tests were found, in general, to be satisfactory.

In another set of experiments carried out in order to find the resistance of the new type of "connector" joints, a somewhat erratic test behaviour of these joints was observed. When taken apart after testing, no defective workmanship in the joints was evident, and the lack of

TIMBER BRIDGES

consistency in their behaviour could not be explained. The strength ratios thus determined by these tests were, however, satisfactory.

4. Plywood I-beams

Another emphatically modern, structural type of timber-work which has developed, particularly in the United States, during the last three decades, is the plywood I-beam.[1]

The advantages of this type are said to be as follows:
(1) These girders can be built of larger cross-sections and lengths than solid beams.
(2) The highly stressed parts can be made of chosen pieces, of particularly high grade.
(3) The relatively small constituent parts are rapidly seasoned.
(4) Plywood beams are easily fabricated and erected.

The specific technique of the type includes several new factors, which are not normally considered in the design of beams made of solid timbers, and deserve, therefore, being men-

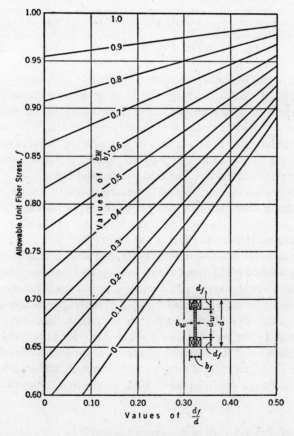

Fig. 74.—Form factors.

[1] For the information on this particular subject, and for the Figs. 74 to 77, the author is indebted to the paper on the "Design of Plywood I-Beams" by howard J. Hansen, published in the *Trans. Am. Soc. C.E.*, Vol. 112, 1947, pp. 955 *et seq.* See also *Modern Timber Design* by the same author (Wiley & Sons, New York, 1943).

tioned even in so short a survey of timber bridges as the present section. Particularly, in regard to the possible future extensions of the scope of application of this new type.

In the first instance attention will be called to the fact that whilst the stress limit sufficed as an almost universal design criterion for all the types hitherto examined, this limit becomes, itself, a function of certain structural dimensions, when a plywood I-beam is designed; namely, as shown in Fig. 74.

This diagram is computed by Mr. Hansen[1] from the formula

$$B = 0.58 + 0.42 \left[u_d \left(1 - \frac{b_w}{b_f} \right) + \frac{b_w}{b_f} \right]$$

which is produced by the Forest Product Laboratory of the United States Department of Agriculture.[2] In this equation B is the "form factor", i.e. the reduction coefficient for the bending stress limit, f, and all other dimensions are as shown in the sketch in the diagram, while u_d depends on the ratio of the depth of the flange to the total depth of the beam, viz. as follows:

d_f/d	0.1	0.2	0.3	0.3	0.5	0.6	0.7	0.8	0.9	1.000
u_d	0.085	0.23	0.40	0.575	0.74	0.875	0.95	0.985	0.998	1.000

The *raison d'être* of this formula is explained by the fact that wood is homogeneous in the statistical sense only, i.e. if we test two large blocks of the same material containing thousands of fibres, we may expect to find their respective resistances to be reasonably similar, but should the sizes of these blocks be reduced to the dimensions of a fibre, the results will depend entirely on whether this happens to be a strong or a weak fibre. Since, therefore, the thicknesses of fibres and plys are of the same order of magnitude, the resistance of a plywood beam calls for a special margin of safety, as compared with a solid beam.

The point of this argument may also be presented in another form, viz. we may say that in a solid beam the strong fibres afford a support to the weak fibres, whereas in the case of a plywood beam, such a compensating action is confined only to the portion of the web, which is protected by the flanges. Hence the effect of the ratio d_f/d.

In the same way as in all other structural types, the design of a plywood beam is guided by its own standards, derived partly from theory and partly from past experience; for instance, since the allowable unit shear over a glued surface is one-eighth only (30 lb/sq in) of the allowable horizontal shear for plywood (240 lb/sq in),[3] the flange-thickness, b_f, cannot be much greater than five times the web thickness, b_w; for otherwise the stress carried by the flange could not be transferred to the web (or *vice versa*).[4] Also, for plywood I-beams unsupported laterally, the maximum ratio of the moment of intertia of the area about the horizontal axis, to its moment of inertia about the vertical axis, must be about 25, and this, for a beam with flange width equal to $5b_w$ and flange depth equal to $10b_w$, means that the unsupported height, d_w, of

---

[1] *Loc. cit.*, p. 957.

[2] "Form Factors of Beams Subjected to Transverse Loading Only", No. 1310, 1941.

[3] These limits decrease as the grade of plywood is reduced below the standard; see: *Technical Data on Plywood*, Douglas Fir Plywood Assn., Tacoma, Wash., U.S.A., 1942.

[4] In the case if instead of glue, the flange is fastened to the web by mechanical means, i.e. bolts or bolts with split rings, the carrying capacity of the flange area must be adjusted to the carrying capacity of these fasteners.

TIMBER BRIDGES

the web must not exceed $9b_w$. It is true that in most bridges the girders are supported laterally by a heavy flooring and by cross-bracings, but since in such works rigidity is generally desirable, it is better not to go far beyond the specified limit for the free height of the web.

Subject to these general desiderata, Hansen gives,[1] in Fig. 75, two sets of curves which yield directly all the basic information required to design a plywood beam; viz. the curves for the section moduli, in the upper part of the diagram, and the curves for the safe load, W, in the lower part thereof.

As seen from the section of the beam shown in this drawing, all the curves appearing therein are calculated for the stipulated proportions of flange width and flange depth to web thickness (viz. $b_w = 0.2 b_f = 0.1 d_f$). The upper curves are supposed to be used in conjunction with Fig. 74, i.e. the permissible stress, by which the section moduli must be multiplied to obtain the bending moments, must be multiplied by the form factor. On the other hand, the lower curves in Fig. 75 are obtained on the assumption that the permissible shearing stress is 240 lb/sq in, and, therefore, if a lower grade of plywood is used, the values yielded by this diagram must be reduced accordingly. Note also that in the case of concentrated loads the values of the diagram must be halved.

There remains one point only: the spacing of the stiffeners. Without going into the details of this problem, which, by the way, is rather more involved than those solved in Figs. 74 and 75, we reproduce in Fig. 76 Mr. Hansen's diagram yielding the required solution for practical application. Note that in this case, in addition to the overall

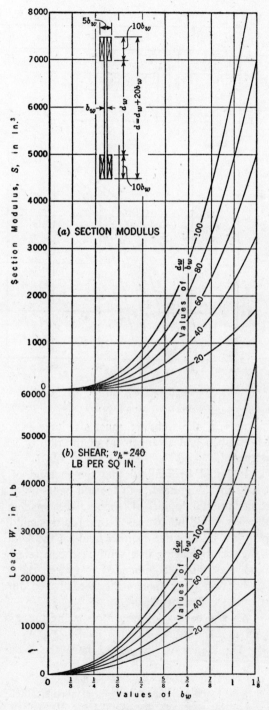

FIG. 75.—Design charts.

[1] *Ibid.*, p. 958.

80 ARCHES AND BRIDGES

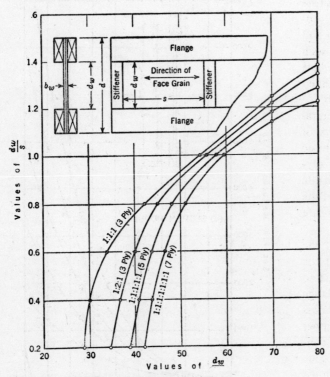

Fig. 76.—Spacing of stiffeners.

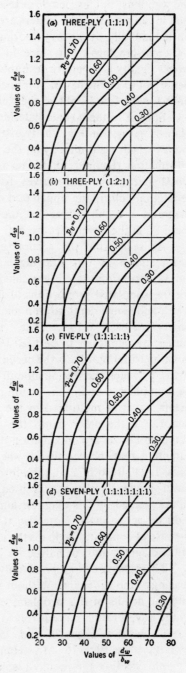

Fig. 77.—Spacing of stiffeners on plywood beams.

thickness of the web, we must also take into account the number and arrangement of the plys of which it is composed.[1]

It should be realized that the method of Fig. 76, strictly speaking, applies to the end panel only, and that from the standpoint of theory, the spacing of the stiffeners could be increased towards the middle of the span, just in the same manner (and for the same reason) as the stirrup spacing in a reinforced concrete beam.[2] It is, however, not recommended to exceed the value $s = 5d_w$, because the stiffeners are needed also during construction, as flange spacers.

An alternative solution for spacing the stiffeners is given by Mr. Sidney Novak[3] in Fig. 77. As compared with Fig. 76, it is alleged to have the advantage of

---

[1] For more ample information on the theoretical argument of Fig. 76 the reader is referred to the original publication.

[2] *Cf.* Mr. Ebeling's discussion of Mr. Hansen's paper, *ibid.*, p. 965.

[3] *Ibid.*, p. 969, Fig. 6.

TIMBER BRIDGES

complying more strictly with the principle of equal factors of safety against horizontal shear and against shear buckling.

The symbol p_v appearing in this diagram is meant to represent the ratio $\dfrac{v_h}{v_{max}/5}$ in which v_h is the design horizontal shear, v_{max} is the maximum horizontal shear stress of the material, and 5 is the theoretical factor of safety against shear buckling. For commercial grades of plywood, the term $v_{max}/5$ should be taken as the maximum horizontal shear stress recommended for that particular grade. On the other hand, it is generally desirable that the value of v_h be kept well below

$$0.83 \times \frac{v_{max}}{5}.$$

This requirement explains why the maximum p_v appearing in the diagram is only 0.70.

(a)

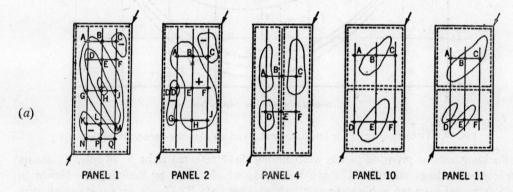

Point[a]	Panel No. 1		Panel No. 2			Panel No. 4	Panel No. 10		Panel No. 11
Load[b]	12,960	14,940	8,950	11,920	14,940	15,920	13,000	18,000	17,700
A	—	$+0\frac{1}{2}$	$+0\frac{3}{8}$	—	$+1\frac{3}{8}$	$+0\frac{1}{2}$	—	—	0
B	—	$+0\frac{3}{16}$	$+0\frac{5}{16}$	—	$+1\frac{1}{16}$	0	$0\frac{5}{8}$	$0\frac{1}{4}$	$+0\frac{1}{4}$
C	—	$+0\frac{1}{8}$	$-0\frac{7}{16}$	—	$+0\frac{1}{16}$	$+0\frac{1}{2}$	—	—	$+0\frac{1}{4}$
D	$+1$	$+1\frac{1}{2}$	$+0\frac{1}{4}$	$+0\frac{5}{8}$	$+0\frac{3}{8}$	$-0\frac{1}{4}$	—	—	$-0\frac{1}{4}$
D'	—	—	—	—	$-0\frac{3}{8}$	—	—	—	—
E	$+0\frac{5}{8}$	$+0\frac{1}{4}$	$+0\frac{1}{2}$	$+0\frac{7}{8}$	$+1\frac{1}{2}$	0	—	1	$+0\frac{1}{4}$
F	$+0\frac{1}{16}$	0	$+0\frac{1}{4}$	$+0\frac{1}{8}$	$+0\frac{7}{8}$	0	$0\frac{7}{8}$	—	0
G	$+0\frac{1}{2}$	$+0\frac{1}{2}$	0	—	$+0\frac{3}{8}$	—	—	—	—
H	$+1\frac{3}{4}$	$+2$	$+0\frac{1}{4}$	—	$+0\frac{5}{16}$	—	—	—	—
J	$+0\frac{11}{16}$	$+0\frac{13}{16}$	$+0\frac{1}{2}$	—	$+1\frac{3}{8}$	—	—	—	—
K	$-0\frac{1}{8}$	$-0\frac{1}{8}$	—	—	—	—	—	—	—
L	$+0\frac{5}{8}$	$+0\frac{5}{8}$	—	—	—	—	—	—	—
M	$+0\frac{7}{8}$	$+1\frac{1}{8}$	—	—	—	—	—	—	—
N	—	$-0\frac{1}{8}$	—	—	—	—	—	—	—
P	—	$-0\frac{1}{8}$	—	—	—	—	—	—	—
Q	—	$+0\frac{5}{8}$	—	—	—	—	—	—	—

(b)

Fig. 78.—(a) Deflections are given for the points shown in the diagrams. (b) Loads on the panel, in pounds, at the time of observation.

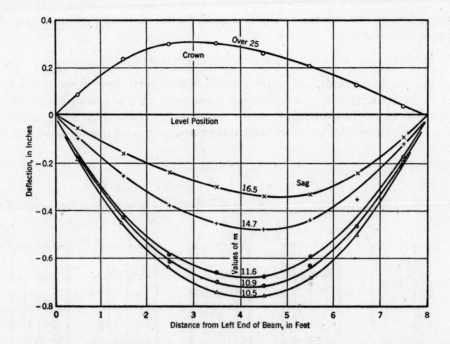

Fig. 79.—Deflection curves for test beam under various moisture-content conditions.

Buckling tests on plywood panels were among those referred to on p. 74 *ante*. Contours of deflection readings taken on the five specimens which failed by buckling are shown in Fig. 78 extracted from the published report[1] about these tests. The loads per panel (which were 8 ft by 4 ft) in pounds and the deflections in inches, corresponding to every contour appearing in the diagram, are given in the table beneath the drawing.

The main conclusion derived from these tests confirmed the general principle that the value of plywood as a diaphragm to carry large loads, even where panels are as thin as 3/8 inch, appears to be established beyond reasonable doubt (*loc. cit.*, p. 685).

It should be realized, however, that although plywood I-beams have now been used in a wide range of applications, the indoor type, including as it does, hangars, store houses, etc., is more frequent than the outdoor type. The bridge engineer might, however, be interested to know also, how a plywood beam would eventually behave, if exposed to air moisture under the conditions of a bridge over a canal. It may, therefore, be pertinent to this discussion to mention that the effect of moisture on the deformation of structural members manufactured in plywood, is capable of systematic study, as witnessed, for instance, by the Wilson and Olson formula:[2]

$$L = L_o \left(1 - \frac{m}{M}\right),$$

which gives the length of plywood, L, at the moisture content m—expressed as the ratio of the

---

[1] "Tests of Timber Structures from Golden Gate International Exposition", by the Committee of the A.S.C.E. of the San Francisco Section (Harold B. Hammill, Chairman) in *Proc. Am. Soc. C.E.*, May 1947, pp. 681 and foll.

[2] W.E. Wilson and Laurence G. Olson, "Deflection of Plywood Beams due to Moisture Content Change", *Proc. Am. Soc. C.E.*, April 1949, Vol. 35, p. 429.

weight of moisture to the oven-dry weight of the plywood. In this formula L_o is the length at zero moisture-content and M is a dimensionless constant which these authors find to be equal to 2.25 (within a certain range of variables). They also show that, using this formula, the deflection of a beam (that may have been originally straight) may be analytically determined, for unequal changes in the moisture contents of its various constituent parts. For instance, Fig. 79 shows one of their experiments[1] with an 8-ft beam with a 6-in web, which agrees reasonably well with the calculated values.

In examining this figure it is essential to realize that the conditions of this experiment exaggerate intentionally the deformations, in order to facilitate measuring them, and thus obtain more precise information on the investigated interdependence between moisture and deformation. In fact the beam tested was different from the standard because:

(a) the flanges were perpendicular to the web;
(b) the web was made of solid timber;
(c) the moisture content of one flange only varied; the other flange and the web remained dry.

In actual out-of-door work the moisture content, if any, would vary gradually from top to bottom of the girder, the effect being, therefore, much less than that observed in the experiment. The diagram must, consequently, be visualized as a matter of general information, but not as a quantitative estimate of the conditions that might occur in Nature.

For temporary bridges (for which the type of beam under consideration seems to be particularly well suited) such deformations are of no consequence. But in permanent work, should the use of the type be extended to such works also, a camber would have to be provided to compensate the estimated effect of the deflection, and thus avoid an unsightly appearance in the future. This, however, applies, as explained earlier, to all types of permanent timber spans, in general.

[1] *Proc. Am. Soc. C.E.*, April 1949, Vol. 35, p. 437.

Chapter Three
ROLLED-JOIST BRIDGES

SECTION I. TWO TYPES

The time-honoured, classical type of rolled-joist bridge was suitable for spans up to twelve metres; but modern rolling-mill technique has produced girders of much greater depths than those in common use in the past, with the result that the scope of application of the type has considerably widened. Such bridges now compete with plate-girder designs; they can be divided into two sub-classes:

(A) Rolled-joist bridges with timber flooring.
(B) „   „   „   „   „ reinforced concrete slab.

SECTION II. TYPE A—ROLLED-JOIST BRIDGES WITH TIMBER FLOORING

1. Pros and Contras
The pros and contras of this type are:

Advantages. (1) The structural weight is low; hence, the masonry piers and abutments are light and may be replaced by trestles.
(2) Minor settlements of the piers or abutments are not dangerous for the safety of the deck.

Disadvantages. (1) Short life of the timber flooring.
(2) Tendency to wear out.

2. Calculations
The flooring is calculated according to the same rules which were given *in extenso* for timber bridges; also, the calculation of the bending moment for the main beams is done in the same manner as explained in the preceding chapter. The section of the rolled joists is then selected from the annexed tables (see pp. 85 to 89) according to the required section modulus.[1]

3. Various Designs
Various designs for fixing the timber flooring on the joists are explained in Figs. 80 to 83. The simplest device is shown in Fig. 80. This arrangement requires a bolt for each plank

---

[1] American engineers provide also for flange buckling and/or web buckling—a procedure seldom (if ever) followed on the older continent, for it is assumed that the extra margin required to cover these contingencies is implicitly included in the adopted stress limit.

Fig. 80.—Type (*A*): Bolts on each plank.

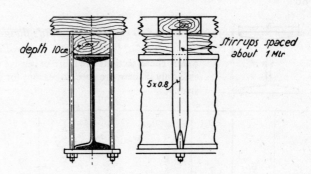

Fig. 81.—Type (*B*): Stirrups.

Standard continental rolled-joist sections

Section No.	Height, mm	Width, mm	Web Thickness, mm	Average Flange Thickness, mm	Sectional Area, cm²	Weight in kg per met	Maximum Permissible Flange Rivet, mm	Moment of Inertia, cm⁴	Section Modulus, cm³	Net Section Modulus after Deduction for Four Rivet Holes, cm³
	h	b	d	t	A	p_d	d_1	I	Z	Z_1
17	170	78	6.6	9.9	25.2	19.78	14	1,166	137	95.3
18	180	82	6.9	10.4	27.9	21.90	do.	1,446	161	114
19	190	86	7.2	10.8	30.6	24.02	do.	1,763	186	134
20	200	90	7.5	11.3	33.5	26.30	17	2,142	214	146
21	210	94	7.8	11.7	36.4	28.57	do.	2,563	244	170
22	220	98	8.1	12.2	39.6	31.09	do.	3,060	278	197
23	230	120	8.4	12.6	42.7	33.52	do.	3,607	314	225
24	240	106	8.7	13.1	46.1	36.19	do.	4,246	354	258
25	250	110	9.0	13.6	49.7	39.01	20	4,966	397	277
26	260	113	9.4	14.1	53.4	41.92	do.	5,744	442	312
27	270	116	9.7	14.7	57.2	44.90	do.	6,626	491	350
28	280	119	10.1	15.2	61.1	47.96	do.	7,587	542	391
29	290	122	10.4	15.7	64.9	50.95	do.	8,636	596	436
30	300	125	10.8	16.2	69.1	54.24	do.	9,800	653	482
32	320	131	11.5	17.3	77.8	61.07	do.	12,510	782	587
34	340	137	12.2	18.3	86.8	68.14	do.	15,695	923	704
36	360	143	13.0	19.5	97.1	76.22	23	19,605	1,089	805
38	380	149	13.7	20.5	107	84.00	do.	24,012	1,264	916
40	400	155	14.4	21.6	118	92.63	do.	29,213	1,461	1,074
42.5	425	163	15.3	23.0	132	103.62	do.	36,973	1,740	1,295
45	450	170	16.2	24.3	147	115.40	26	45,852	1,937	1,579
47.5	475	178	17.1	25.6	163	127.96	do.	56,481	2,378	1,817
50	500	185	18.0	27.0	180	141.30	do.	67,738	2,750	2,127
55	550	200	19.0	30.0	213	167.21	do.	90,184	3,607	2,886
60	600	215	21.6	32.4	254	199.40	do.	138,957	4,632	3,735

ARCHES AND BRIDGES

Wide flange continental rolled joists

Note: These sections are more expensive and are recommended for use in the case only if there are specific reasons for reducing the structural height of the span.

Section No.	Height, mm	Width, mm	Web Thickness, mm	Flange Thickness, mm	Sectional Area, cm²	Weight in kg per met	Maximum Permissible Flange Rivet, mm	Moment of Inertia, cm⁴	Section Modulus, cm³	Net Section Modulus after Deduction for Four/Two Flange Rivets, cm³
	h	b	d	t	A	p_d	d_1	I	Z	Z_1
20	200	200	8.5	18.12 / 9.5	70.4	55.3	23	5,171	517	409
25	250	250	10.5	21.70 / 10.9	105.1	82.5	do.	12,066	965	789
30	300	300	12.5	26.25 / 13.25	152.1	119.4	26	25,201	1,680	1,391
40	400	do.	15.5	31.0 / 18.2	203.6	159.8	do.	57,834	2,892	2,414
50	500	do.	19.4	35.2 / 22.6	261.8	205.5	do.	111,283	4,451	3,747
60	600	do.	20.8	37.2 / 24.7	300.6	236.0	do.	179,303	5,977	5,074
70	700	do.	21.1	37.5 / 25.0	325.2	255.3	do.	258,106	7,374	6,245
80	800	do.	21.5	38.5 / 26.0	354.9	278.6	do.	360,486	9,012	7,744
90	900	do.	do.	do.	376.4	295.5	do.	473,964	10,533	9,092
100	1,000	do.	21.9	39.5 / 21.9	407.2	319.7	do.	621,287	12,425	10,769

Standard continental channels

14	140	60	7.0	10.0	20.4	16.01	17	605	86.4	65.9
16	160	65	7.5	10.5	24.0	18.84	20	925	116	86.3
18	180	70	8.0	11.0	28.0	21.98	do.	1,354	150	116
20	200	75	8.5	11.5	32.2	25.28	do.	1,911	191	150
22	220	80	9.0	12.5	37.4	29.36	23	2,690	245	179
24	240	85	9.5	13.0	42.3	33.21	do.	3,598	300	236
26	260	90	10.0	14.0	48.3	37.92	do.	4,823	371	296
28	280	95	do.	15.0	53.3	41.84	do.	6,276	448	362
30	300	100	do.	16.0	58.8	46.16	26	8,026	535	425

ROLLED-JOIST BRIDGES

American standard rolled-joist sections

Note: These are the shapes of the Association of American Steel Manufacturers, but each steel company has also its own list of special sections which can be found in their books of tables.

Nominal Size, in	Weight, lb/ft	Sectional Area, in^2	Depth of Beam, inches	Width of Flange, inches	Web Thickness, inch	Average Flange Thickness, inch	Max. Flange Rivet, inch	Moment of Inertia, in^4	Section Modulus, in^3
			h	b	d		d_1	I	Z
$6 \times 3\frac{3}{8}$	17.25	5.02	6	3.565	0.359	0.359	$\frac{5}{8}$	26.0	8.7
$7 \times 3\frac{1}{2}$	15.3	4.43	7	3.660	0.250	0.392	do.	36.2	10.4
do.	20.0	5.83	do.	3.860	0.450	do.	do.	41.9	12.0
8×4	18.4	5.34	8	4.000	0.270	0.425	$\frac{3}{4}$	56.9	14.2
do.	23.0	6.71	do.	4.171	0.441	do.	do.	64.2	16.0
$10 \times 4\frac{5}{8}$	25.4	7.38	10	4.660	0.310	0.491	do.	122.1	24.4
do.	35.0	10.22	do.	4.944	0.594	do.	do.	145.8	29.2
12×5	31.8	9.26	12	5.000	0.350	0.544	do.	215.8	36.0
do.	35.0	10.20	do.	5.078	0.428	do.	do.	227.0	37.8
$12 \times 5\frac{1}{4}$	40.8	11.84	do.	5.250	0.460	0.659	do.	268.9	44.8
do.	50.0	14.57	do.	5.477	0.687	do.	do.	301.6	50.3
$15 \times 5\frac{1}{2}$	42.9	12.49	15	5.500	0.410	0.622	do.	441.8	58.9
do.	50.0	14.59	do.	5.640	0.550	do.	do.	481.1	64.2
18×6	54.7	15.94	18	6.000	0.460	0.691	$\frac{7}{8}$	795.5	88.4
do.	70.0	20.46	do.	6.251	0.711	do.	do.	917.5	101.9
$20 \times 6\frac{1}{4}$	65.4	19.08	20	6.250	0.500	0.789	do.	1,169.5	116.9
do.	75.0	21.90	do.	6.391	0.641	do.	do.	1,263.5	126.3
20×7	85.0	24.80	do.	7.053	0.653	0.916	1	1,501.7	150.2
do.	95.0	27.74	do.	7.200	0.800	do.	do.	1,599.7	160.0
24×7	79.9	23.33	27	7.000	0.500	0.871	do.	2,087.2	173.9
do.	90.0	26.30	do.	7.124	0.624	do.	do.	2,230.1	185.8
do.	100.0	29.25	do.	7.247	0.747	do.	do.	2,371.8	197.6
$24 \times 7\frac{7}{8}$	105.9	30.98	do.	7.875	0.625	1.102	do.	2,811.5	234.3
do.	120.0	35.13	do.	8.048	0.798	do.	do.	3,010.8	250.9

and for each girder. A more satisfactory solution is embodied in Fig. 81. A longitudinal timber is placed over each joist and is secured to the metal by stirrups. The planks of the flooring are then nailed to these auxiliary timbers. This design reduces the shock (impact) transmitted to the steel beam.

Longitudinal timbers are also used in Figs. 82 and 83, but here they rest on the transverse ties. On the other hand, Fig. 84, including as it does a cast-iron pipe, is an older type.

Transverse ties, as shown in Figs. 82 to 84, are spaced about 2 metres apart (5 to 6 feet). In addition to this, diagonal ties, usually made of 3×3-inch angle-bars or similar sections, are riveted or bolted to the lower flange of the joists, at about 45° to the axis of the span.

ARCHES AND BRIDGES

BRITISH STANDARD SECTIONS
BEAMS

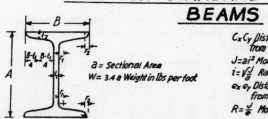
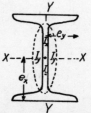

a = Sectional Area
$W = 3.4a$ Weight in lbs per foot

$C_x C_y$ Distance of Centre of Gravity from X axis and Y axis.
$J = ai^2$ Moment of Inertia.
$i = \sqrt{\frac{J}{a}}$ Radius of Gyration.
$e_x e_y$ Distance of outer fibres from X and Y axes.
$R = \frac{J}{e}$ Moment of Resistance.

2	3	4	5	6	7		8		11	12	11₁	12₁	15	16	15₁	16₁	17
SIZE	STANDARD THICKNESS		RADII		WEIGHT		SECTIONAL AREA		MOMENTS OF INERTIA				MOMENTS OF RESISTANCE				BSB
$A \times B$	t_1	t_2	r_1	r_2	W		a		I_x	I_y	I_x	I_y	R_x	R_y	R_x	R_y	No.
Inches	Inches				lbs/ft	Kg/Mts	Inches²	Cm²	Inches⁴		Cm⁴		Inches³		Cm³		
3 × 1½	.160	.248	.260	.150	4.00	5.952	1.176	7.587	1.657	.124	68.969	5.161	1.105	.165	18.108	2.704	1
3 × 3	.200	.332	.300	.150	8.50	1.265	2.501	16.135	3.789	1.261	157.710	52.487	2.526	.841	41.394	13.781	2
4 × 1¾	.170	.240	.270	.135	5.00	7.440	1.472	9.497	3.671	.194	152.798	8.075	1.835	.222	30.070	3.638	3
4 × 3	.220	.336	.320	.160	9.50	14.136	2.795	18.032	7.526	1.280	313.255	53.277	3.763	.854	61.664	13.994	4
4¾ × 1¾	.180	.325	.280	.140	6.50	9.672	1.912	12.335	6.767	.263	281.663	10.947	2.849	.300	46.687	4.916	5
5 × 3	.220	.376	.320	.160	11.01	16.383	3.238	20.890	13.620	1.461	566.905	60.811	5.448	.974	89.276	15.961	6
5 × 4½	.290	.448	.390	.195	17.99	26.769	5.290	34.129	22.699	5.656	944.800	235.420	9.080	2.514	148.794	41.197	7
6 × 3	.260	.348	.360	.180	11.99	17.841	3.527	22.755	20.228	1.338	841.950	55.692	6.743	.892	110.498	14.617	8
6 × 4½	.370	.431	.470	.235	20.00	29.760	5.882	37.948	34.660	5.409	1442.653	225.139	11.553	2.404	189.319	39.394	9
6 × 5	.410	.520	.510	.255	25.00	37.200	7.354	47.445	43.641	9.105	1816.469	378.977	14.547	3.642	238.382	59.681	10
7 × 4	.250	.387	.350	.175	16.01	23.823	4.709	30.381	39.222	3.410	1632.537	141.934	11.206	1.705	183.633	27.940	11
8 × 4	.280	.402	.380	.190	18.01	26.799	5.297	34.174	55.716	3.574	2319.067	148.761	13.929	1.787	228.255	29.284	12
8 × 5	.350	.575	.450	.225	28.02	41.694	8.241	53.168	89.357	10.250	3719.306	426.606	22.339	4.100	366.069	61.187	13
8 × 6	.440	.597	.540	.270	35.00	52.080	10.293	66.406	110.597	17.929	4603.379	746.259	27.649	5.976	453.084	97.929	14
9 × 4	.300	.460	.400	.200	21.00	31.248	6.178	39.858	81.115	4.198	3376.250	174.733	18.026	2.099	295.392	34.396	15
9 × 7	.550	.924	.650	.325	58.02	86.334	17.064	110.090	229.740	46.265	9562.468	1925.688	51.053	13.219	836.606	216.620	16
10 × 5	.360	.552	.460	.230	29.99	44.625	8.820	56.903	145.684	9.780	6063.805	407.073	29.137	3.912	477.468	64.106	17
10 × 6	.400	.736	.500	.250	42.02	62.526	12.358	79.729	211.614	22.930	8808.010	954.415	42.323	7.643	693.547	125.246	18
10 × 8	.600	.970	.700	.350	69.98	104.130	20.582	132.787	345.039	71.609	14361.558	2980.581	69.008	17.902	1130.834	293.360	19
12 × 5	.350	.550	.450	.225	31.99	47.601	9.408	60.697	220.115	9.743	9161.847	405.533	36.686	3.897	601.173	63.860	20
12 × 6	.400	.717	.500	.250	44.02	65.502	12.946	83.522	315.439	22.257	13129.517	926.403	52.573	7.419	861.514	121.575	21
12 × 6	.500	.883	.600	.300	53.99	80.337	15.879	102.445	375.599	28.280	15633.557	1177.098	62.600	9.427	1025.826	154.480	22
14 × 6	.400	.698	.500	.250	46.01	68.463	13.533	87.310	440.625	21.584	18340.134	898.391	62.946	7.197	1031.496	117.904	23
14 × 6	.500	.873	.600	.300	57.01	84.831	16.769	108.187	533.091	27.941	22188.847	1162.988	76.156	9.314	1247.983	152.629	24
15 × 5	.420	.647	.520	.260	41.09	61.142	12.351	79.684	428.207	11.937	17823.260	496.854	57.094	4.775	935.599	78.248	25
15 × 6	.500	.880	.600	.300	58.98	87.762	17.346	111.909	629.094	28.203	26184.780	1173.893	83.879	9.401	1374.525	154.054	26
15 × 6	.550	.847	.650	.325	61.97	92.211	18.227	117.583	725.953	27.069	30216.342	1126.693	90.744	9.023	1487.022	147.860	27
18 × 7	.550	.928	.650	.325	75.02	111.630	22.066	142.361	1149.667	46.618	47852.590	1940.361	127.741	13.320	2093.292	218.275	28
20 × 7½	.600	1.010	.700	.350	88.96	132.372	26.164	168.800	1671.291	62.586	69564.145	2605.017	167.129	16.690	2738.743	273.499	29
24 × 7½	.600	1.070	.700	.350	99.93	148.696	29.392	189.625	2654.769	66.874	110499.430	2783.497	221.123	17.833	3625.312	292.229	30

ROLLED-JOIST BRIDGES

BRITISH STANDARD SECTIONS
CHANNELS

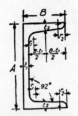

a = Sectional Area
$W = 3.4a$ Weight in lbs. per foot.

C_x, C_y Distance of centre of gravity from X axis & backline of channel.
$I = ai^2$ Moment of Inertia.
$i = \sqrt{\frac{I}{a}}$ Radius of Gyration.
e_x, e_y Distances of outer fibres from X and Y axes.
$R = \frac{I}{e}$ = Moment of resistance

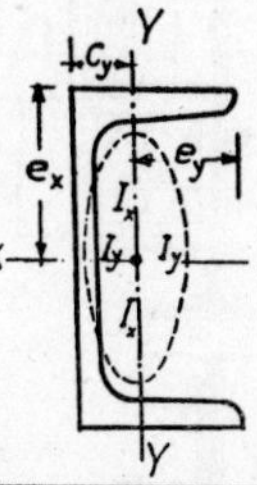

2	3	4	5	6	7		8		11	12	11,	12,	15	16.	15,	16,	17
SIZE	STANDARD THICKNESS	RADII			WEIGHT W		SECTIONAL AREA a		MOMENTS OF INERTIA				MOMENTS OF RESISTANCE				BSC
$A \times B$	t_1	t_2	r_1	r_2					I_x	I_y	I_x	I_y	R_x	R_y	R_x	R_y	N°
Inches	Inches	Inches	Inches	Inches	lbs/foot	kg/MT	Inches²	Cm²	Inches⁴	Inches⁴	Cm⁴	Cm⁴	Inches³	Inches³	Cm³	Cm³	
3 × 1½	·250	·312	·312	·220	5.27	7.842	1.549	9.994	1.994	·296	82.996	12.320	1.329	·291	21.778	4.769	1
3½ × 2	·250	·312	·312	·220	6.75	10.044	1.986	12.813	3.701	·713	154.047	29.677	2.115	·526	34.659	8.620	2
4 × 2	·250	·375	·375	·260	7.96	11.844	2.341	15.103	5.709	·843	237.626	35.088	2.855	·627	46.785	10.275	3
5 × 2½	·312	·375	·375	·260	10.98	16.338	3.230	20.839	12.134	1.774	505.053	73.835	4.854	1.018	79.542	16.682	4
6 × 2½	·312	·375	·375	·260	12.04	17.916	3.542	22.852	18.763	1.880	780.972	78.251	6.254	1.047	102.484	17.157	5
6 × 3	·312	·437	·437	·300	14.49	21.561	4.261	27.490	24.010	3.503	999.368	145.805	8.003	1.699	131.145	27.842	6
6 × 3	·375	·475	·475	·325	16.29	24.240	4.791	30.910	26.034	3.822	1083.613	159.083	8.678	1.845	142.206	30.234	7
6 × 3½	·375	·475	·475	·325	17.90	26.635	5.266	33.974	29.656	5.907	1234.372	245.867	9.885	2.481	161.985	40.656	8
7 × 3	·375	·475	·475	·325	17.56	26.129	5.166	33.329	37.627	4.017	1566.149	167.200	10.751	1.889	176.177	30.955	9
7 × 3½	·400	·500	·500	·350	20.23	30.102	5.950	38.387	44.549	6.498	1854.263	270.466	12.728	2.664	208.574	43.655	10
8 × 2½	·312	·437	·437	·300	15.12	22.499	4.448	28.697	41.094	2.283	1710.456	95.025	10.273	1.245	168.344	20.402	11
8 × 3	·375	·500	·500	·350	19.30	28.718	5.675	36.613	53.432	4.329	2224.000	180.186	13.358	2.008	218.898	32.905	12
8 × 3½	·425	·525	·525	·375	22.72	33.807	6.682	43.110	63.763	7.067	2654.007	294.150	15.941	2.839	261.225	46.523	13
8 × 4	·450	·530	·550	·375	25.73	38.286	7.569	48.832	74.018	10.790	3080.851	449.112	18.504	3.855	303.225	63.172	14
9 × 3	·375	·437	·437	·350	19.37	28.823	5.696	36.748	65.177	4.021	2712.862	167.366	14.484	1.790	237.349	29.333	15
9 × 3½	·375	·500	·500	·350	22.27	33.137	6.550	42.258	79.902	6.963	3325.761	289.821	17.756	2.759	290.968	45.212	16
9 × 3½	·450	·550	·550	·375	25.39	37.780	7.469	48.187	88.075	7.660	3665.946	318.832	19.572	3.029	320.726	49.636	17
9 × 4	·475	·575	·575	·400	28.55	42.482	8.396	54.168	101.654	11.635	4231.144	484.284	22.590	4.084	370.182	66.925	18
10 × 3½	·375	·500	·500	·350	23.55	35.042	6.925	44.677	102.622	7.187	4271.436	299.145	20.524	2.800	336.327	45.804	19
10 × 3½	·475	·575	·575	·400	28.21	41.976	8.296	53.522	117.959	8.194	4909.807	341.059	23.592	3.192	386.602	52.307	20
10 × 4	·475	·575	·575	·400	30.16	44.878	8.871	57.232	130.715	12.018	5440.750	500.225	26.143	4.147	428.405	67.957	21
11 × 3½	·475	·575	·575	·400	29.82	44.372	8.771	56.587	148.606	8.421	6185.428	350.507	27.019	3.234	442.760	52.996	22
11 × 4	·500	·600	·600	·425	33.22	49.431	9.771	63.039	170.454	12.812	7094.807	533.274	30.992	4.362	507.866	71.480	23
12 × 3½	·375	·500	·500	·350	26.10	38.837	7.675	49.516	158.639	7.572	6603.031	315.169	26.440	2.868	433.272	46.998	24
12 × 3½	·500	·600	·600	·425	32.88	48.925	9.671	62.393	190.135	8.922	7958.963	371.360	31.789	3.389	520.926	55.536	25
12 × 4	·525	·625	·625	·425	36.47	54.267	10.727	69.206	218.181	13.654	9081.348	568.320	36.363	4.599	595.880	75.364	26
15 × 4	·525	·630	·630	·440	41.94	62.407	12.334	79.574	377.007	14.554	15692.162	605.781	50.268	4.748	823.742	77.805	27

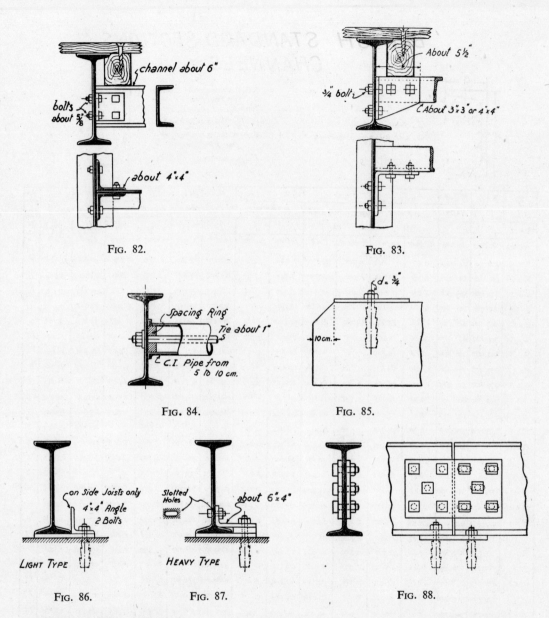

Fig. 82. Fig. 83.

Fig. 84. Fig. 85.

Fig. 86. Fig. 87. Fig. 88.

4. Bearings and Base-plates

Ashlar or concrete bearing blocks, about 50 cm (20 inches) in width, are usually provided on piers and abutments to support the joists. They carry base-plates, which may either be continuous steel plates, about $\frac{5}{8}$ inch thick, or independent cast-iron bearings, provided specially under each joist. A margin of about 10 cm (4 inches) is left from the edge of the plate to the face of block (see Fig. 85). The joists are fixed (on the base-plates) in such a manner that a free movement, due either to elastic deformation or to thermal expansion, can take place, without causing the beam to twist (see Figs. 86 to 88).

SECTION III. TYPE B—ROLLED-JOIST BRIDGES WITH REINFORCED-CONCRETE SLAB

1. Pros and Contras

This was at one time a very popular type both for railway and road traffic. In modern applications of this class of bridge, concrete and steel co-operate both, in resisting the longitudinal bending moment.

Advantages. (1) Longer life, as compared with a timber decking.
(2) The formwork for casting the concrete can be attached to the joists themselves; and consequently, there is no obstruction to the flow of water under the bridge, whilst carrying out the work.
(3) Lighter than reinforced concrete bridges and less dependent on settlements of abutments and piers.
(4) The span may be supported on steel trestles.

Disadvantages. Under normal conditions a span formed wholly in reinforced concrete is more economical, because the difference in the volume of concrete does not counterbalance the extra expenses involved in the excess of steel.

2. Calculations

Where a rolled-joist bridge with timber flooring exists, and the flooring is to be renewed, it may be preferable to replace it by a reinforced concrete slab, thus obtaining a better class of bridge; however, if an entirely new bridge is to be built, local conditions and considerations of permanency will dictate the choice of the type, and the materials to be adopted.

3. Three Subclasses

Rolled-joist bridges with reinforced concrete slabs can be subdivided into several subclasses, as follows:

Type I.—The reinforced concrete constitutes a decking only, and is placed above the steel.

Type II.—The joists are embedded in concrete; in this case they may either carry all the load by themselves or may be calculated as the reinforcement of the concrete.

Type III.—As type I, but the slab participates in resisting the main bending moment.

In the design of the first type the calculation of the joists does not call for any special remark. The maximum bending moment is figured out in the usual way as explained for timber bridges in Chapter Two. Knowing then the permissible bending stress, it is easy to find the required section modulus and, using the attached tables of rolled sections, select the depth and type of joist answering the specific requirements of the case in hand.

SECTION IV. CALCULATION OF SLAB

1. General Order of Procedure

This is possibly the most conventional of all the civil engineering calculations discussed in these pages.

It is supposed that the weight on the wheel is carried by a strip of the slab, of a certain width l (see Fig. 89), the remainder of the slab being thus completely disregarded.

2. "French" Formula

The value of l had been repeatedly determined, experimentally and theoretically; but the results thus arrived at might be subject to different interpretations, and consequently empirical equations which practice has shown to be safe and reliable are commonly preferred in design practice; however, since the arbitrary element plays in this case a predominant part, the designer cannot be said to have mastered the subject until he has compared the relative merits of a certain number of such empirical equations. In carrying out this investigation we shall choose,

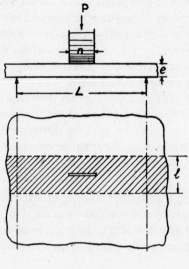

Fig. 89.

as the first instance, the so-called "French" formula, which in spite of its old age—it was first included in the French Ministerial Circular of 1906—is still most popular in several countries. This formula reads

$$l = \frac{L}{3} + e$$

in which L is the span and e the slab thickness plus paving.

Comparing this equation with other empirical formulae, it will easily be found that it is rather conservative; in applying this formula it will therefore be permissible to make liberal assumptions with regard to other points of stress analysis, taking into account the superabundant safety of the equation. In fact French engineers frequently reduce the value of the calculated bending moment in order to obtain a more economical result; for instance, they replace in calculating this moment the effective span by the clear opening. When applying this formula, it must be remembered that the concentrated load is assumed to be distributed over a rectangle, the area of which is

$$(e + L/3) \times e.$$

It is supposed that the secondary, distributary reinforcement suffices to ensure such a distribution. For that purpose it must be half of the main reinforcement.

3. "American" Formula

For our second example we shall use an American formula, viz. the Ketchum's formula, which is

$$l = \frac{2L}{3} + n$$

where n is the width of the wheel, for slabs supported on transverse girders.

For girder bridges this is reduced to $l = \frac{2L}{3}$.

4. "German" Method

For identification purposes in the following discussion we may describe the two quoted formulae as embodying respectively the French and American principles. We shall then find that the prevailing tendency in Germany is to use the American rule for the middle of the span, but to employ the French rule at the supports, i.e. to calculate the bending stresses in the middle of the span assuming the effective width as $\frac{2L_1}{3}$, but to take one third of L only (plus twice the thickness of the slab) for the calculation of the shear at the supports.

On the whole this way of proceeding is not unreasonable, for a simple analysis of the lines of forces supplies evidence confirming the general principle that a wider part of the slab participates in supporting a concentrated load placed in the middle of the span, rather than when the load is at the end of the span. On the other hand it may be objected that in so simple a case, an attempt to rationalize an obviously arbitrary rule might not be worth the trouble, *le jeu ne vaut pas la chandelle*. This, however, is a matter of opinion.

5. Comparison and Example of Application

The application of these formulae in practice will now be explained in an example.

The same case will be considered as for timber bridges, but it will be supposed that the deck consists of a concrete slab resting on rolled joists.

It will be remembered (see p. 63) that the live-load moment was found to be

$$M_e = 11{,}850 \text{ kg} \times \text{metres}.$$

For the structural weight we will assume $p_d = 1{,}000$ kg per metre of the joist. Hence:

$$M_d = \frac{p_d L^2}{8} = 1{,}000 \; \frac{7.6^2}{8} = 7{,}220 \text{ kg} \times \text{metres}.$$

$$M_{total} = 11{,}850 + 7{,}220 = 19{,}070 \text{ kg} \times \text{metres}.$$

Assuming $f = 1{,}000$ kg/cm², we require $Z = \frac{1{,}907{,}000}{1{,}000} = 1{,}907$ cm³.

Select from table on p. 88 joist B.S.B.—28—

Dimensions: $18'' \times 7''$. Section Modulus $Z = 127.74$ in³ $= 2{,}000$ cm³. Weight 75.02 lb/ft.

We shall take the same number of joists as for the wooden bridge, namely five. Spacing, 130 cm centre to centre.

In applying the French formula the span is measured between the outer edges of the

---

[1] Provided this does not exceed the width of the wheel, plus twice the paving, plus 2 metres (6 feet).

beams, or joists; as the flange of the selected section is 7 inches wide and the joists are spaced 130 cm centre to centre, the span for our slab will be

$$130 - (7 \times 2.5) = 112.5 \text{ cm.}$$

The thickness, e, which includes slab and paving, is not constant, owing to the camber in the middle of the roadway, but it may be assumed about 20 cm, as an average value. Since the width of the wheel is 50 cm and the thickness 20 cm, the load is distributed over 70 cm (see Fig. 90).

It follows that the live bending moment is

$$M_e = \frac{6,000}{2} \left(\frac{1.125}{2} - \frac{0.70}{4} \right) = 1,162.5 \text{ kg} \times \text{metres.}$$

The effect of continuity of the slab may here be represented by the coefficient $K = 0.85$. The effective width will then be

$$l = 1/3 \times 1.125 + 0.20 = 0.575 \text{ metre.}[1]$$

Hence, the live-load moment *per metre width of the slab* is

$$M'_e = \frac{M_e \times K}{l} = 1,718.5 \text{ kg} \times \text{metres.}$$

The structural dead weight per square metre is about 530 kg for the reinforced concrete slab and 150 kg for the asphalt-brick paving. Thus the total structural weight per square metre is 680 kg.

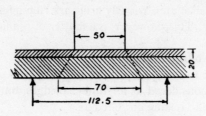

Fig. 90.

Consequently

$$M_d = \frac{680 \times 1.125^2}{8} = 107.61 \text{ kg} \times \text{metres.}$$

The effect of continuity is now $K_1 = \frac{8}{10}$.

$$M'_d = 8/10 \times 107.61 = 86.09 \text{ kg} \times \text{metres.}$$

Hence: $M_{total} = 1,804.59 \text{ kg} \times \text{metres.}$

6. Required Thickness of Slab

The next point is to find the required thickness of the slab.

The fundamental principle of reinforced concrete design in civil engineering, is that for beams and slabs in bending, the theoretical thickness of the concrete and the sectional area of

---

[1] Note that this width will also appear later, in the calculation of the shear.

TABLE OF COEFFICIENTS

FOR CALCULATING EFFECTIVE DEPTH OF SLAB OR BEAM FROM SQUARE ROOT OF BENDING MOMENT; METRIC UNITS; RATIO $n = \frac{E_s}{E_c} = 12$

Formula $d = K_1 \sqrt{\frac{M}{b}}$, where d = effective depth in cm, K_1 = coefficient appearing in table, M = bend. moment in kgs. × met, and b = width, in metres.

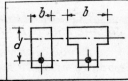

f_s stress in steel kgs./cm²	f_c = stress in concrete in kgs. per sq. cm.									
	30	35	40	45	50	60	70	80	90	100
	Values of coefficient K_1									
800	0.489	0.433	0.390	0.357	0.330	0.289	0.259	0.237	0.219	0.204
900	0.508	0.448	0.403	0.368	0.340	0.297	0.266	0.242	0.223	0.208
1000	0.526	0.463	0.416	0.379	0.349	0.304	0.272	0.247	0.227	0.212
1100	0.543	0.477	0.428	0.389	0.358	0.312	0.278	0.255	0.232	0.215
1200	0.559	0.491	0.440	0.400	0.367	0.319	0.284	0.257	0.236	0.219

(A)

TABLE OF COEFFICIENTS

FOR CALCULATING EFFECTIVE DEPTH OF SLAB OR BEAM FROM SQUARE ROOT OF BENDING MOMENT; METRIC UNITS; RATIO $n = \frac{E_s}{E_c} = 15$

Formula $d = K_1 \sqrt{\frac{M}{b}}$, where d = effective depth in cm, K_1 = coefficient appearing in table, M = bend. moment in kgs. × met, and b = width, in metres.

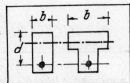

f_s stress in steel kgs./cm²	f_c = stress in concrete in kgs. per sq. cm.									
	30	35	40	45	50	60	70	80	90	100
	Values of coefficient K_1									
800	0.459	0.408	0.369	0.339	0.314	0.276	0.249	0.228	0.212	0.198
900	0.474	0.420	0.380	0.348	0.322	0.283	0.254	0.232	0.215	0.201
1000	0.489	0.433	0.390	0.357	0.330	0.289	0.259	0.237	0.219	0.204
1100	0.504	0.445	0.401	0.366	0.338	0.295	0.264	0.241	0.222	0.207
1200	0.519	0.457	0.411	0.375	0.345	0.301	0.269	0.245	0.226	0.210

(B)

TABLE OF COEFFICIENTS
FOR CALCULATING AREA OF TENSILE REINFORCEMENT FROM SQUARE ROOT OF BENDING MOMENT; METRIC UNITS; RATIO $n = \frac{E_s}{E_c} = 12$

Formula $a = K_2\sqrt{Mb}$, where a = area of bars in cm²; K_2 = coefficient appearing in table, M = bend. moment in kgs. x met. and b = width, in metres.

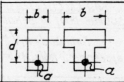

f_s stress in steel kgs./cm²	f_c = stress in concrete in kgs. per sq. cm.										
	30	35	40	45	50	60	70	80	90	100
	Values of coefficient K_2										
800	0.285	0.326	0.366	0.403	0.442	0.514	0.580	0.647	0.707	0.765
900	0.242	0.277	0.312	0.345	0.377	0.440	0.500	0.554	0.609	0.659
1000	0.209	0.240	0.270	0.299	0.327	0.383	0.435	0.484	0.531	0.579
1100	0.182	0.210	0.236	0.262	0.287	0.337	0.384	0.431	0.471	0.510
1200	0.161	0.186	0.209	0.233	0.255	0.299	0.341	0.380	0.420	0.456

Note: This table must be used in the case only if the depth, d, is calculated from the corresponding table on p. 95.

(C)

TABLE OF COEFFICIENTS
FOR CALCULATING AREA OF TENSILE REINFORCEMENT FROM SQUARE ROOT OF BENDING MOMENT; METRIC UNITS; RATIO $n = \frac{E_s}{E_c} = 15$

Formula $a = K_2\sqrt{Mb}$, where a = area of bars in cm²; K_2 = coefficient appearing in table, M = bend. moment in kgs. x met. and b = width, in metres.

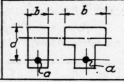

f_s stress in steel kgs./cm²	f_c = stress in concrete in kgs. per sq. cm										
	30	35	40	45	50	60	70	80	90	100
	Values of coefficient K_2										
800	0.310	0.354	0.395	0.437	0.475	0.549	0.618	0.684	0.748	0.808
900	0.264	0.301	0.338	0.372	0.406	0.478	0.531	0.589	0.645	0.697
1000	0.228	0.261	0.293	0.324	0.353	0.410	0.464	0.517	0.567	0.612
1100	0.200	0.228	0.257	0.285	0.311	0.363	0.409	0.458	0.500	0.542
1200	0.177	0.203	0.229	0.253	0.276	0.322	0.366	0.409	0.450	0.485

Note: This table must be used in the case only if the depth, d, is calculated from the corresponding table on p. 95.

(D)

TABLE OF COEFFICIENTS
FOR CALCULATING EFFECTIVE DEPTH OF SLAB OR BEAM FROM SQUARE ROOT OF BENDING MOMENT; INCH POUND UNITS; RATIO $n = \frac{E_s}{E_c} = 12$

Formula $d = K_1 \sqrt{\frac{M}{b}}$, where d = effective depth in inch, K_1 = coefficient appearing in table, M = bend. moment in lbs × inch, and b = width in inches.

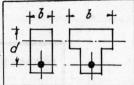

f_s stress in steel lbs./inch	f_c = stress in concrete in lbs per sq. inch.						
	500	550	600	650	700	800	900
	Values of coefficient K_1						
12,000	0.117	0.108	0.101	0.095	0.090	0.081	0.075
14,000	0.122	0.113	0.105	0.099	0.093	0.084	0.077
15,000	0.124	0.115	0.107	0.101	0.095	0.086	0.079
16,000	0.127	0.118	0.110	0.103	0.097	0.087	0.080
17,000	0.130	0.120	0.112	0.105	0.099	0.089	0.081
20,000	0.137	0.127	0.118	0.110	0.104	0.093	0.085

(E)

TABLE OF COEFFICIENTS
FOR CALCULATING EFFECTIVE DEPTH OF SLAB OR BEAM FROM SQUARE ROOT OF BENDING MOMENT; INCH. POUND UNITS; RATIO $n = \frac{E_s}{E_c} = 15$

Formula $d = K_1 \sqrt{\frac{M}{b}}$, where d = effective depth in inch, K_1 = coefficient appearing in table, M = bend. moment in lbs × inch, and b = width in inches.

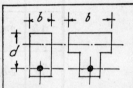

f_s stress in steel lbs./inch.	f_c = stress in concrete in lbs per sq. inch.						
	500	550	600	650	700	750	800
	Values of coefficient K_1						
12,000	0.109	0.102	0.095	0.090	0.085	0.081	0.077
14,000	0.114	0.106	0.099	0.093	0.088	0.084	0.080
15,000	0.116	0.108	0.101	0.095	0.090	0.085	0.081
16,000	0.118	0.110	0.103	0.096	0.091	0.086	0.083
17,000	0.121	0.111	0.104	0.098	0.093	0.088	0.084
20,000	0.127	0.118	0.109	0.103	0.097	0.092	0.088

(F)

Round Reinforcement Bars Commonly Used

DIAMETER		AREA		PERIMETER		WEIGHT	
Inches	cms.	Square Inches	Square Cms.	Inches	cms.	Lbs. per foot	kgs. per mt.
1/4"	0.635	0.0491	0.3168	0.785	1.994	0.167	0.249
5/16"	0.794	0.0767	0.4948	0.982	2.494	0.261	0.388
3/8"	0.952	0.1104	0.7122	1.178	2.992	0.376	0.560
1/2"	1.270	0.1963	1.2664	1.571	3.990	0.668	0.994
5/8"	1.587	0.3068	1.9793	1.963	4.986	1.043	1.552
3/4"	1.905	0.4418	2.8502	2.356	5.984	1.502	2.235
7/8"	2.222	0.6013	3.8792	2.749	6.982	2.044	3.042
1"	2.540	0.7854	5.0669	3.142	7.981	2.670	3.973
1 1/8"	2.857	0.9940	6.4127	3.534	8.976	3.380	5.030
1 1/4"	3.175	1.2272	7.9171	3.927	9.974	4.172	6.209

(G)

the steel are both roughly proportional to the square root of the bending moment, i.e. this principle satisfies the requirement that the stresses in both steel and concrete are the maximum permissible stresses. It is true that some authors recommend to use the cubic root of the moment, particularly in building construction, as the more economical basis for proportioning reinforced concrete in bending, but this does not apply to our case.

Tables (A) to (G) contain numerical coefficients to be used in this calculation, both for English and metric units; these coefficients depend on the permissible stress limits and value of $n = \frac{E_s}{E_c}$, i.e. on the ratio of the Young Moduli for steel and concrete respectively. In practice, such tables are always employed. The trial-and-error method (starting with a certain assumed depth-to-span ratio) has outlived its time and is no longer used in design.

ROLLED-JOIST BRIDGES

We shall now adopt $n=15$ and assume for the stresses in concrete and steel respectively the limits:

$$f_c = 50 \text{ kg/cm}^2$$
$$f_s = 1{,}000 \text{ ,, ,,}$$

The tables B and D on pp. 95 and 96 will then yield the following coefficients: 0.33 and 0.35.

Including, therefore, these values into the "square-root" formulae we obtain for the theoretical thickness of the slab and for the reinforcement area:

$h = 0.33 \sqrt{M_{total}} = 14$ cm.

$a = 0.35 \sqrt{M_{total}} = 14.88$ cm². We assume 8 bars $\tfrac{5}{8}$ inch, per metre, which gives an area of $8 \times 1.98 = 15.84$ cm². In carrying out this calculation it is recommended that table G be used, giving the sectional areas of round bars.

Allowing 3.0 cm between centre of bar and bottom of slab, the total thickness is $14.0 + 3.0 = 17$ cm. The design is shown in Fig. 91.

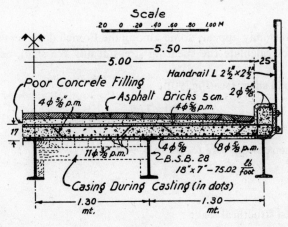

Fig. 91.

7. Checking the Stress

The percentage of steel (this conventional percentage is calculated for the theoretical thickness of the slab only, and only for the tensile reinforcement) is found to be:

$$p = \frac{8 \times 1.98}{14.0 \times 100} = 0.0113.$$

We assume as before $n = 15$; hence

$$np = 15 \times 0.0133 = 0.1695.$$

Let y be the distance to the neutral axis and $x = \dfrac{y}{h}$. Then, according to the usual formula:

$$x = -np + \sqrt{(np)^2 + 2np} = 0.423$$

and
$$y = xh = 0.423 \times 14 = 5.92 \text{ cm.}$$

The second moment is therefore equal to

$$I = \frac{y^3 \times 100}{3} + (h-y)^2 \times 8 \times 1.98xn = 22{,}429 \text{ cm}^4.$$

From this we obtain the stresses:

$$f_c = \frac{M}{I} y = \frac{180{,}500}{22{,}429} \, 5.92 = 47.86 \text{ kg/cm}^2;$$

$$f_s = \frac{M}{I} (h-y) n = \frac{180{,}500}{22{,}429} \, 8.08 \times 15 = 975 \text{ kg/cm}^2,$$

which is near enough to the assumed figures.

We shall later give and explain (in the chapter on Reinforced Concrete Bridges) the use of diagrams facilitating the calculation of stresses in reinforced concrete slabs and beams. But when a single calculation only is to be performed, as in the case in hand, it is preferable to apply the simple arithmetical solution, as given above, because under such conditions it takes more time to understand the method of the diagram than to perform the more laborious arithmetic of the elementary theory.

8. The Shear

Strange though this may appear, some among the French designers in applying their formula for slabs, do not calculate the shearing stress at all. We shall, nevertheless, carry out this verification.

Maximum live-load shear (see Fig. 90):

$$S_e = \frac{6{,}000 \times 77.5}{112.5} = 4{,}130 \text{ kg.}$$

Live shear per metre width (see footnote p. 94):

$$S'_e = \frac{4{,}130}{0.575} = 7{,}180 \text{ kg.}$$

Dead shear due to structural load:

$$S_d = 680 \, \frac{112.5}{2} = 382 \sim 380 \text{ kg.}$$

Total shear per metre width $= S_{total} = 7{,}180 + 380 = 7{,}560$ kg/metre.

Arm of internal couple $= h - 1/3y = 14 - \frac{5.92}{3} = 12.03$ cm.

Shearing stress $= \frac{7{,}560}{12.03 \times 100} = 6.28$ kg/cm².

As this is greater than the permissible 4 kg/cm², we shall bend 50 per cent of the bars at about one-fifth of the span.

In addition to which, 4 bars $\frac{3}{8}$ inch in diameter will be placed at the top of the slab, to resist the negative moment.

The longitudinal reinforcement, which is the secondary reinforcement in this case, will consist of 11 bars $\frac{3}{8}$ inch diameter, giving a total area $17 \times 0.712 = 7.84$ cm², viz. about half the area of the main reinforcement (see "French" formula).

9. Application of Different Formulae

Let us now check the stresses in the slab which we have designed, by applying the "American" formula referred to earlier.

Since the true span, as measured between the axes of the joists, is now introduced into the calculation, the live moment will now be equal to:

$$M_e = \frac{6,000}{2}\left(\frac{1.30}{2} - \frac{0.70}{4}\right) = 1,425 \text{ kg} \times \text{metres}.$$

The "effective width" will be determined as follows:

$$l = 2/3 \times 1.30 = 0.87 \text{ metre}.$$

Consequently, per metre width of slab, the live moment will be

$$M'_e = \frac{1,425 \times 0.85}{0.87} = 1,391 \text{ kg} \times \text{metre}.$$

The corresponding dead moment is

$$M_d = \frac{680 \times 1.30^2}{10} = 115 \text{ kg} \times \text{metres}.$$

We thus find the total moment:

$$M_{total} = M_e + M_d = 1,506 \text{ kg} \times \text{metres}.$$

This yields the following stresses:

$$f_c = \frac{M}{I} y = \frac{150,600 \times 5.92}{22,429} = 39.7 \text{ kg/cm}^2.$$

$$f_s = \frac{M}{I}(h-y)n = \frac{150,600 \times 8.08}{22,429} = 812.10 \text{ kg/cm}^2.$$

These values are about 20 per cent less than those previously deduced from the French formula, which confirms the conservative character of the French method as compared with analogous equations used in other countries.

Similar results will be arrived at from the "German" formula; at least in so far as the centre of the span is concerned. However, if we were to check the shearing stresses at the support according to the German method, a greater thickness would be found necessary, which explains that haunches are rather popular in that country.

10. Conclusions

The quantitative comparison of three different calculation methods shows a surprisingly wide divergence in the calculated results, which seems to be symptomatic of the effect of the arbitrary assumptions, playing a predominant part in these various solutions.

There are good reasons for believing that this state of affairs is, at least partly, due to the fact that the slabs of current proportions are generally capable of carrying greater loads than those calculated from the usual equations. From structural considerations we cannot use a thinner slab, because it would then be impossible (or, possibly, difficult) to carry out the job properly by means of the usual constructional methods—but, once this requirement is satisfied, we obtain a slab of greater strength than given by the formula. The latter reflects, therefore, the practical requirements rather than the theoretical stress limits.

SECTION V. STRUCTURAL DETAILS OF SLAB

1. General Remark

All the information appearing in this paragraph will also apply (so far as the context admits) to the purely reinforced concrete spans discussed in the next section.

2. Ends of Bars

The ends of a reinforcing bar are usually bent to form hooks. These hooks were first suggested by Considère. The classical experiments, carried out in Germany by Professor Bach, yielded the well-known table given below.

Type of bars	Breaking load	Sketch
Bars without hooks	100 per cent	
Bars with rectangular hooks	152 ,, ,,	
Bars with U-shaped hooks	153 ,, ,,	
Bars with U-hooks and transverse bars passing through the hooks	160 ,, ,,	

The failure of the beam in the upper sketch was due to the sliding of bars in the concrete; in all other cases it was the crushing effect of the concrete in the hook.

Reinforcing bars are therefore usually provided with hooks, and whenever possible a transverse bar is arranged in each hook. Revolutionary tendencies aiming at saving the cost of hooks, and maintaining that this could be done without material effect on structural resistance, have sporadically made their appearance among builders and designers, but the profession *as a whole* does not appear to have followed in their wake.

3. Junction of Bars

It is often difficult, or even impossible, to find bars sufficiently long to cover the full length of the bridge, and it is therefore necessary to make them out of several shorter pieces.

In contract drawings those junctions are usually not shown, as they depend on the steel available on the market at the moment of erection; but the form and size of the junctions must be given explicity in the specification accompanying the design, for they depend upon the permissible stresses in steel and concrete.

The force carried by a bar is transmitted to another bar by the concrete in between them. Assuming the bars to overlap, the force transmitted is, obviously, proportional to the length of the overlap. The next problem is therefore, to find the minimum length, l, of the overlap. With reference to the notation in Fig. 92, the tensile force in the bar is

$$\frac{\pi d^2 f_s}{4}$$

where f_s is the tensile limit for the steel.

On the other hand, the force transmitted from the bar to the concrete is equal to $\pi d l f_a$, where f_a is the permissible adhesion stress between steel and concrete.

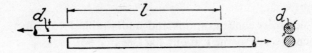

Fig. 92.

The following condition must therefore be satisfied

$$\frac{\pi d^2 f_s}{4} = \pi d l f_a.$$

It follows that

$$l = \frac{d}{4} \frac{f_s}{f_a}.$$

Thus, for $f_s = 1,000$ kg/cm² and $f_a = 4.0$ kg/cm², we have $l = 63d$.

This length may be reduced by using hooks at the ends of the bars. Experiments are supposed to have shown that, in this case, the rupture takes place by shear in the concrete, the surface of rupture being ab (see Fig. 93). Hence, the area subjected to shear is:

$$5d \times c + \frac{(5d)^2 \pi}{4} + c\pi d,$$

in which c is the length of the straight portion of the hook (see figure).

The equation is therefore as follows:

$$f_a \left(25 \frac{\pi d^2}{4} + 5cd + c\pi d \right) = f_s \frac{\pi d^2}{4}$$

from which

$$c = \frac{0.7854 f_s/f_a - 19.63}{8.14} d.$$

The total length of the hook, inclusive of the straight part, is $l = c + 2d + \frac{5}{2} \pi d$.

Thus, for $f_s = 1000$ kg/cm² and $f_a = 4.0$ kg/cm², we find $l = 31.6d$.

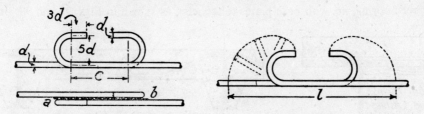

Fig. 93.

In order to compare this arrangement with the first alternative (straight bars), it will suffice to calculate the total length of the overlapping steel (see Fig. 93). This will be:

$$c + 6d + 5\pi d = 53.3d.$$

This is less than $63d$, and it follows, therefore, that the second alternative (bars with hooks) is more economical. This is the traditional argument substantiating the method usually adopted.

4. Bends

In order to avoid stress concentrations and undesirable local effects, the radius of the bend must be sufficiently great, namely about $20d$ (see Fig. 94).

5. Under-surface of Slab

The bottom of the concrete slab in a rolled-joist bridge may either be flush with the top of the upper flange of the joist, or it may be set at about the same level as the lower face of this flange (see Fig. 95). The second method seems to be more popular in America; the lower reinforcement of the slab is then laid directly on the flange of the joist. The advantage of this method lies in the constructional feasibility of the casing which is required for casting the concrete (see dotted lines in Fig. 91).

Various other arrangements are also frequently used, as for instance that shown in Fig. 96. For moderate-span bridges such as those discussed in these pages, considerations relating to the economy of erection methods, are usually as important as the economy of the main work.

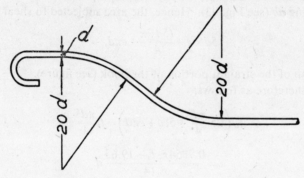

Fig. 94.

6. Bearings

The bearings of the joists may be designed in the same manner as for rolled-joist bridges with timber flooring, i.e. with base-plates (channels), as shown in Figs. 97 and 98; or, alterna-

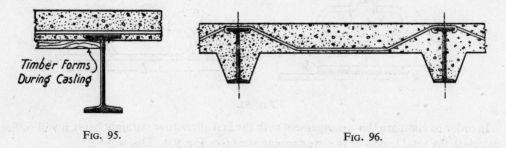

Fig. 95.   Fig. 96.

tively, with the spaces in between the joists filled on the abutments with concrete, thus forming end-bearing blocks, as shown in Fig. 99.

ROLLED-JOIST BRIDGES

7. Bracings

The cross-bracings may either be formed of angle bars, about 3×3 inches, riveted or bolted to the lower flange of the main joists; or of concrete cross-girders, cast together, and forming T-beams with the slab. In the first case the slab is calculated as if supported on two sides

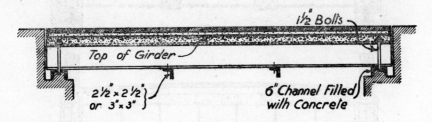

FIG. 97.

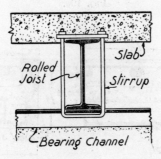

FIG. 98.

only, whereas with the second arrangement the designer may assume effective support on all four sides.

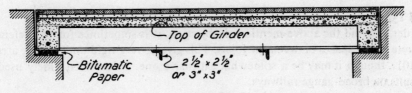

FIG. 99.

The second type of device is illustrated in Fig. 100.

In this case the handrails are preferably made of concrete and the extreme or outside girders completely embedded. Special holes are then drilled in the joists in order to accommodate the reinforcement of the cross-girders.

The fact that temperature variations have a different effect upon embedded and exposed metal, respectively, is the greatest drawback to this type. It may lead, in hot climates, to awkward temperature cracks.

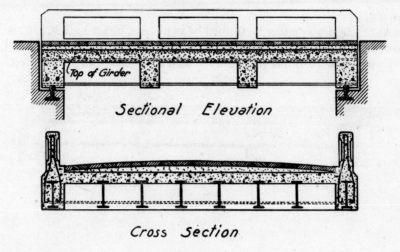

Fig. 100.

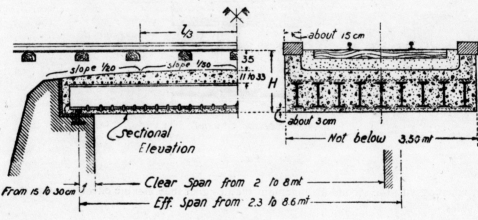

Fig. 101.

8. Special Types

In order to avoid the above-mentioned objection, it is sometimes found preferable to fill with concrete all the spaces between the joists. This gives a very rigid and durable arrangement (see Fig. 101), though it may be a somewhat expensive one. It has been widely used with excellent results on broad-gauge railways.

In this case, owing to the great depth of concrete, the usual numerical computation of its strength (in resisting the bending effect) is dispensed with; but it may be necessary to verify that the tensile stress in the concrete does not exceed 20 kg/cm^2, assuming the usual railway load to be distributed over 3.5 metres width. The 20 kg limit is believed to represent the moment when the concrete begins to detach itself from the steel on the underface of the bridge.

According to the standards of the Berlin Railway, as shown in Fig. 101, the total depth of such bridges is given by the formulae:

(a) For spans less than 6 metres $H = 0.62 + 0.1L$ metre.
(b) For spans equal to or greater than 6 metres $H = 0.81 + 0.065L$ metre.

ROLLED-JOIST BRIDGES

Fig. 102 illustrates a lighter type belonging to the same group of works.

SECTION VI. ROLLED-JOIST-BRIDGES, IN WHICH STEEL AND CONCRETE PARTICIPATE IN CARRYING BENDING MOMENT

1. Introductory

This class of designs forms a hybrid type, being partly a steel bridge and partly a reinforced concrete span. The following notation is adopted (see Fig. 103):

h is the distance from the top of the slab to the bottom of the joist.
e_1 is the thickness of the web of the joist.
e_2 is the mean thickness of the flange of the joist.
h_s is the height of the joist.
A_s is the sectional area of the joist $= h_s e_1 + 2e_2 (B - e_1)$.
I_s is the second moment of the joist $= \dfrac{e_1 h_s^3}{12} + 2 \left(\dfrac{h_s - e_2}{2} \right)^2 (B - e_1) e_2$.

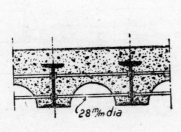

Fig. 102.

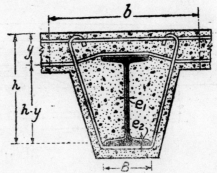

Fig. 103.

The equation of moments is:

$$\tfrac{1}{2} b y^2 - n A_s \left(h - \frac{h_s}{2} - y \right) = 0.$$

From this we determine $y = \dfrac{nA_s}{b} \left[\sqrt{1 + \dfrac{(2h - h_s) b}{nA_s}} - 1 \right].$

Thereafter the second moment is found from

$$I = \tfrac{1}{3} b y^3 + n I_s + n A_s \left(h - \frac{h_s}{2} - y \right)^2.$$

The stresses in concrete (f_c) and in steel (f_s) are then:

$$f_c = \frac{M}{I} y \quad \text{and} \quad f_s = \frac{M}{I} n (h - y).$$

2. Earlier Designs versus Modern Tendencies

In earlier designs of this type (see Figs. 102 and 103) the steel girder was entirely embedded in the concrete. Modern tendencies allow departures from this earlier rule. For instance, in Fig. 104 the top part only of the girder is surrounded by concrete, whereas in Fig. 105 the slab

108   ARCHES AND BRIDGES

is altogether above the girder.[1] In the latter case, however, in order to develop the required resistance against shear, parts of another girder, forming steel blocks, are welded to the main girder. This method is used also in Fig. 106, whereas in Fig. 107 a spiral wire is welded on the top of the joists according to the patented system Alpha (Switzerland). The last two figures are prepared for the particular purpose to explain graphically the intended objective of the steel devices welded, according to this new technique, to the top flange of the joists and embedded in the concrete slab.

In this connection attention is called to the small sketches of beams with exaggerated deflections, as they appear in the last two drawings. They show that, had the slab and the joist

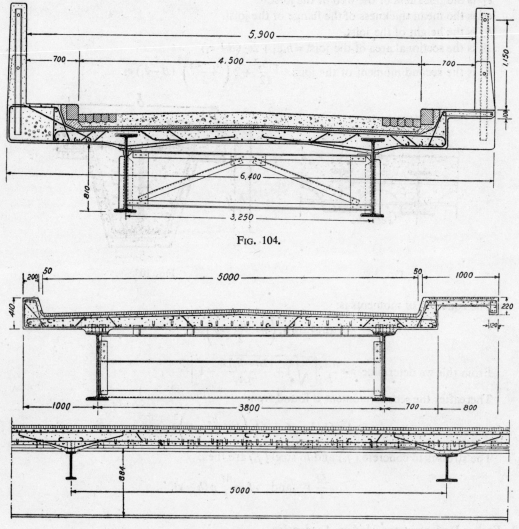

Fig. 104.

Fig. 105.

[1] From *Träger in Verbund-Bauweise* by the Technical Commission of the Union of Swiss Bridge Builders and Steel Construction Firms, Professor Dr. M. Roš, President, Zürich, March 1944 (in French and German).

ROLLED-JOIST BRIDGES

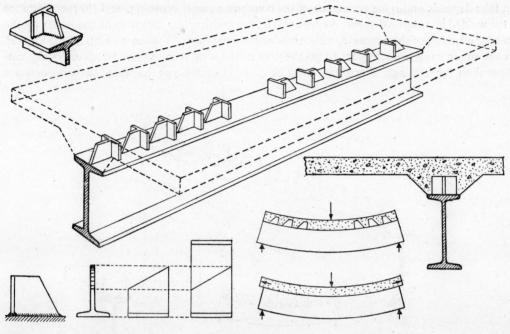

Fig. 106.

been left to expand independently, the one would have slipped over the other, as shown in the lower sketches—the relative movement being zero in the centre and rising to a maximum at either end. Hence, a vertical plane *drawn through both slab and joist* would not have remained plane after the deflection had taken place. This would have prevented us from using the moment equation, as given above for the built-up girder, inclusive of both steel and concrete.

3. Shear "Connectors"

The object of the steel elements welded to the joists and encased by the concrete of the slab (also called "shear connectors" required for adequate bond at the steel-concrete-interface) is to ensure vertical planes remaining plane after deflection, as shown in the upper sketches in the same figures. In precisely the same manner and for the same object as the keys in the built-up timber girder, which we have designed earlier (see p. 64 to 68). In fact these keys were intended to prevent the upper wooden beam from slipping over the lower wooden beam. It might, therefore, be of interest to compare the deflection sketches in Figs. 106 and 107 with Figs. 59 and 60. The analogy will then be obvious.

It follows, consequently, that the calculation of the required strength, and the number, of the blocks welded on top of the joist in Fig. 106 must be done according to the shearing stress diagram, in precisely the same way as was explained for the keys in pages 65 to 67. Subject, however, to the obvious exception, that the individual force that each block can carry is a specific problem, different from that of timber keys dealt with earlier.

It will be observed, in the first instance, that every block, as shown in Fig. 106, is cut out from a steel joist and welded to the top surface of the main steel beam. Since such a block acts as a buffer arresting the potential movement of the concrete material, the critical force it

110 ARCHES AND BRIDGES

can take depends on (a) the resistance of the concrete against crushing, and (b) the resistance of the welded joint against shear. As regards (a), we multiply the face area of the buffer by the limit for crushing of the concrete, which, under circumstances, is taken as high as 125 kg/cm². On the other hand, with respect to (b) the area of the weld is multiplied by about 750 kg/cm² (shear limit for welding). The design must be made in such a way that these two products are

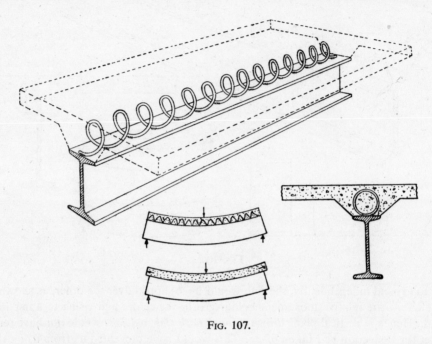

FIG. 107.

equal, and the spacing of the blocks is then computed as for the keys, substituting this product for the individual resistance of a key.

4. Patented Continuous "Connectors"

Similar considerations may be proffered in regard to the patented shear-connector system shown in Fig. 107, with that difference, however, that instead of the concentrated type of shear resistances, it affords a continuous shear resistance diagram.

Note that in this type of bridge, metal can be economized by using two different sizes of joists, cutting them lengthwise into two parts each, and welding the parts of the lighter profiles over the parts cut from the heavier sections. In this manner the neutral axis of the compound girder can be made to approach the point at mid-height of the web, which is always an advantage.

It may also be of interest to mention that in order to eliminate the effect of the shrinkage of concrete and the ensuing possibility of cracks, Professor F. Dischinger, of the Technical University of Berlin, suggests[1] pre-stressing the reinforced slabs in the built-up girders described above. In fact, before deflection takes place, shrinkage may cause undesirable tensile stresses in the slab, and this can be avoided by pre-stressing.

[1] F. Dischinger, "Stahlbrücken in Verbund mit Stahlbetondruckplatten bei gleichzeitiger Vorspannung durch hochwertige Seile", *Bauingenieur* (1949), pp. 321, 364.

Chapter Four
REINFORCED CONCRETE BRIDGES

SECTION I. GENERAL PRINCIPLES

1. T-beam Bridges Only

We will consider in these pages T-beam bridges only. It will be realized that the application of the writer's abridged calculation method for slab and T-beam reinforced concrete bridges, which is presented in the chapter on regulators (see the author's *Irrigation Engineering*, Vol. I of *Design Textbooks in Civil Engineering*, pp. 172–180, Chapman and Hall, Ltd.), is conditional on the same restrictive considerations which are usually valid for all empirical solutions in general; i.e. though the results of this empirical method are fully satisfactory, it does not constitute a self-contained theory, for it is derived from a number of existing designs, prepared by the rational method which is described below. Thus, unless the argument of the rational procedure is fully mastered by the designer, he must follow blindly the empirical rules given earlier, without understanding what they actually mean. It is therefore essential in his own interest that he should read carefully and assimilate the discussion which follows.

2. Reader Supposed to be Acquainted with General Principles of R.C. Design

On the other hand, it should be understood that this is not a course on reinforced concrete. The reader is supposed to be acquainted with the general principles of this subject; but the author's experience, in schoolroom and designing office, shows that such general knowledge does not suffice for the preparation of a feasible project of even a short span bridge, because there are a variety of *sui generis* methods and rules, which constitute bridge engineering as such, and do not belong to other types of design. An attempt at these rules is presented below.

3. Limits of Spans for Various Types

In the common type of reinforced concrete road bridge of the slab type, such for instance as that standardized on the crossings of Egyptian roads with irrigation canals, the depth of the slab must not exceed 26 cm (10 inches); after this limit is exceeded, it usually becomes more economical to provide ribs, in order to subdivide the span of the slab and thus reduce its thickness. We must obtain the T-beam bridge.

Should the 20-ton standard type of lorry or the 60-ton trailer, be adopted as live load (see Figs. 53 to 55), it will be found, as a corollary to the above rule, that slab bridges are economically sound for spans up to 3 metres (10 feet). For greater spans, the slab must be supported on T-beams, and, in this case, its depth will vary from 14 to 20 cm ($5\frac{1}{2}$ to 8 inches).

ARCHES AND BRIDGES

4. Steel Percentages

The percentage of steel is greater in T-beam bridges than in slabs. The average ratio of steel to concrete in some of the newly built bridges is as follows:

(*a*) For slab bridges: 95 kg per cubic metre of concrete.
(*b*) For T-beam bridges: 120 kg per cubic metre of concrete.

The extreme limit for the simply supported T-beam type of reinforced concrete bridge is a span of 11 to 13 metres (35 to 40 feet). Above this limit the structural weight becomes excessive, thus rendering the design uneconomical. Greater spans (20 to 40 per cent longer) can be covered by cantilever-beam bridges of the same type.

SECTION II. EXAMPLE OF R.C. BRIDGE. DESIGN OF SLAB

1. General Notes

The routine followed in designing a reinforced concrete bridge can best be explained by means of an example.

The same case as before will be considered here, namely, a 7-metre wide bridge, designed for a 7-metre clear span and a 20-ton standard lorry.

The bridge consists of a 5-metre roadway and two footpaths, 1 metre-wide each. The roadway is paved with compressed asphalt bricks on a bedding of poor concrete of variable

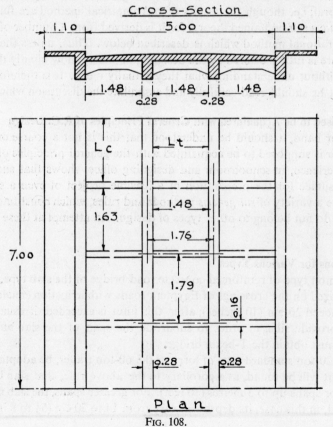

Fig. 108.

REINFORCED CONCRETE BRIDGES

thickness, arranged in such a way as to provide the required transverse camber.

The design is begun by preparing a sketch of the bridge (see Fig. 108) giving the preliminary layout of beams and slabs. As the calculation proceeds, this layout is gradually amended to conform with the results obtained.

The purpose of this preliminary sketch is to form an approximate estimate of the respective spans of the different constituent parts of the reinforced concrete decking. It follows that the plan, in which these spans are clearly shown, represents the significant part of the drawing, the sections serving as an illustration only. In drawing such plans for the type here considered we may assume that the width of the ribs of the main girders varies, depending on the span, from 25 cm (10 inches) to 35 cm (14 inches). In the same way, the range of widths convenient for cross-girders will be 14 to 18 cm ($5\frac{1}{2}$ to 7 inches). These data will enable the student to draw the preliminary layout of the bridge, without necessarily referring to other existing works.

The number of main beams is in our case taken to be four, namely: two intermediate and two extreme beams. It is usually found convenient for bridges of usual width, to arrange the spacing of these beams in such a way that the clear distance between them should be of the order of about 1.5 metres (5 feet). For instance, had the width of the bridge been 5 metres, we would have used three beams, and so on.

2. Spacing Cross-girders

The next point is to space the cross-girders properly. This is done in such a manner as to obtain, as nearly as possible, a square panel. Assuming the width of the main girders to be about 28 cm (see Fig. 108), the transverse span of the slab is $\frac{5.0 - 2 \times 0.28}{3} = \frac{4.44}{3} = 1.48$ metres. Taking 16 cm for the width of the cross-girder, we shall find:

(1) Should three cross-girders be used, the longitudinal span of the slab would be:

$$\frac{7.0 - 3 \times 0.16}{4} = 1.63 \text{ metres.}$$

(2) For four cross-girders, it would come to:

$$\frac{7.0 - 4 \times 0.16}{5} = 1.27 \text{ metres.}$$

Consequently, three cross-girders yield a panel as nearly square as feasible. Hence, this

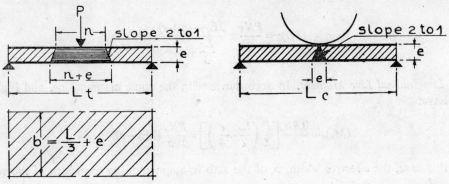

Fig. 109.

solution is adopted, and the panel of the slab is therefore 1.48×1.63 metres, or $L_T = 1.48$ metres, and $L_e = 1.63$ metres. We will assume the depth of the slab to be 16 cm (for the calculation of the weight only; the final figure will be the result of the stress analysis, which follows).

The dead weight of the slab will then be found as follows:

Reinforced concrete $1.0 \times 0.16 \times 2,400 = 384$ kg/sq metre
Asphalt bricks and poor concrete $= 180$,, ,, ,,
Total p_d $\overline{564}$,, ,, ,,

In round figures, 570 kg/metre².

3. Calculation of Moments for Design of Slab

In calculating the bending moments for the slab, we will use the following formulae:

(a) *Dead-load Moment*. The structural weight of reinforced concrete is here taken to be 2,400 kg/cubic metre. The effect of continuity is assumed to be equivalent to $K_1 = 0.8$; wherefrom the formula:

$$M_d = 0.8 p_d N_e \frac{L^2}{8} = N_e \frac{p_d L^2}{10},$$

in which N_e is the coefficient of reduction, due to the fact that the slab is supported on four sides (and not on two sides, as would have been the case with a girder bridge).

(b) *Transverse Live Moment*. Here again the continuity is assumed to be equivalent to $K_1 = 0.8$. The effective width, b, is one-third of the span, plus the thickness of the slab (according to the "French" formula, see p. 92). Consequently (see Fig. 109):

$$(M_e)_T = \frac{0.8}{b} N_T \left[\frac{P}{2} \left(\frac{L_T}{2} - \frac{n+e}{4} \right) \right] = \frac{PN_T}{10b} [2L_T - (n+e)]$$

where P is the concentrated wheel load, N_T is the coefficient representing the transverse effect of the reduction due to the four supports and n is the width of the wheel.

Substituting in this

$$b = \frac{L_T}{3} + e$$

we find

$$(M_e)_T = \frac{PN_T}{10} \times \frac{2L_T - (n+e)}{\frac{L_T}{3} + e}.$$

(c) *Longitudinal Live Moment*. In accordance with the same assumptions, and Fig. 109, we will have:

$$(M_e)_e = \frac{0.8 N_e}{b} \left[\frac{P}{2} \left(\frac{L_e}{2} - \frac{e}{4} \right) \right] = \frac{PN_e}{10b} (2L_e - e).$$

In this case, the effective width, b, of the slab is augmented by the width of the wheel; hence $b = \frac{L_e}{3} + e + n$

REINFORCED CONCRETE BRIDGES

$$(M_e)_e = \frac{PN_e}{10} \times \frac{2L_e - e}{\frac{L_e}{3} + e + n}.$$

(d) *Negative Dead-weight Moment over the Support.* This is assumed equal to the positive dead-weight moment, as for a continuous beam over three spans.

(e) *Negative Live Moment.* At the same point, is also taken to be equal to the corresponding positive moment, as it would be under the combined action of two wheels of the same lorry, placed in adjacent spans.

Using, now, these formulae, and assuming $e = 0.16 + 0.06 = 0.22$ metre (depth of slab, plus paving, plus depth of poor concrete), the following numerical results are arrived at:

Positive Moments:
Transverse.

$$(M_d)_T = \frac{N_T p_d L_T^2}{10} = \frac{0.4 \times 570 \times 1.48^2}{10} = 50 \text{ kg} \times \text{metres}.$$

$$(M_e)_T = \frac{PN_T}{10} \times \frac{2L_T - (n+e)}{\frac{L_T}{3} + e} = \frac{6{,}000 \times 0.4}{10} \times \frac{2 \times 1.48 - (0.50 + 0.22)}{\frac{1.48}{3} + 0.22}$$

$$= 757 \text{ kg} \times \text{metres}.$$

Total transverse moment $M_T = 757 + 50 = 807$ kg × metres.

4. Formulae for Slabs Supported on Four Sides

The coefficient N_T included in these formulae is calculated from another "French" formula:

$$N_T = \frac{L_e^4}{L_e^4 + L_T^2 L_e^2 + L_T^4} = \frac{1.63^4}{1.63^4 + 1.48^2 \times 1.63^2 + 1.48^4} = \frac{7.07}{17.73} = 0.3987.$$

The diagram in Fig. 110 gives, in a simple graphical form, a number of alternative equations which have been suggested by various authors in lieu of this formula.[1] The values of the coefficients, for various values of a and b, may be scaled off the diagram. Particular attention is called to the average formula derived graphically by the author, viz.

$$K = \frac{1}{1 + 1.6 (a/b)^2 \sqrt{a/b}}$$

in which $K = N$.

5. Final Values for Bending Moments and Corresponding Design of Slab
Longitudinal Moments:

$$(M_d)_e = \frac{N_e p_d L_e^2}{10} = \frac{0.273 \times 570 \times 2.66}{10} = 41 \text{ kg} \times \text{metres}.$$

$$(M_e)_e = \frac{PN_e}{10} \times \frac{2L_e - e}{\frac{L_e}{3} + e + n} = \frac{6{,}000 \times 0.273}{10} \times \frac{2 \times 1.63 - 0.22}{\frac{1.63}{3} + 0.72}$$

$$= 394 \text{ kg} \times \text{metres}.$$

---

[1] It was first used by the author when drafting the original General Specification for the Bridges of the Egyptian State Railways.

in which
$$N_e = \frac{L_T^4}{L_e^4 + L_T^2 L_e^2 + L_T^4} = \frac{4.84}{17.73} = 0.273.$$

Thus, the total longitudinal moment $M_e = 394 + 41 = 435$ kg × metres.

Transverse Reinforcement. This in our case is the main reinforcement, because the transverse moment, M_T, is greater than the longitudinal moment, M_e.

Assuming, then,
$$f_c = 35 \text{ kg/cm}^2.$$
$$f_s = 900 \text{ ,, \quad ,,}$$
$$n = \frac{E_s}{E_c} = 12$$

we find $\quad d_e = 0.448 \sqrt{807} = 0.448 \times 28.41 = 12.72$ cm

in which 0.448 is the coefficient appearing in the fifth line of the third column in the table A on page 95 against the values $f_c = 35$ kg/cm² and $f_s = 900$ kg/cm².

The total thickness of the slab is therefore 16 cm (in accordance with the original assumption). The required area of reinforcing bars is:

$$a_s = 0.277 \sqrt{807} = 0.277 \times 28.41 = 7.87 \text{ cm}^2;$$

in which 0.277 is the corresponding coefficient read off the table C on p. 96. We then assume 11 bars per metre, $\frac{3}{8}$ inch diameter.

According to p. 98 the actual area is $11 \times 0.712 = 7.83$ cm².

Every alternate bar will be bent, and $5\frac{1}{2}$ bars per metre, $\frac{3}{8}$ inch diameter, will be used as upper reinforcement of the slab.

Longitudinal Reinforcement. This must at least be half the main reinforcement (see p. 92 *ante*); we therefore assume 8 bars per metre $\frac{3}{8}$ inch diameter at the bottom of the slab, and as much on the top. Generally speaking, for bridges supporting heavy concentrated live loads, such as are here considered, 8 bars per metre is a minimum (5 inches centre to centre) for slab reinforcement. Also, under usual conditions 12 bars per metre are a maximum, for reasons pertaining to execution.[1]

6. Adopting Low Stress Limits ($f_0 = 35$ kg/cm² and $f_s = 900$ kg/cm²)

Modern specifications for buildings and other works which are not designed for rolling loads, allow much higher stresses, but for bridges an impact percentage is frequently included. Since for the type of span here considered the dynamic effect is practically the same for every part of its anatomy, there is no material advantage in introducing the impact addition, for the same result can be obtained, using a constant, reasonably low stress limit for the entire work.

It must also be realized that the weight of the live loads circulating on modern roads has increased during the last thirty years almost in the same proportion as the permissible stress. Thus, the standard 6-ton wheel-load, with a reasonably conservative stress limit, still remains the correct combination.

---

[1] The number of bars per unit width of reinforced concrete slab is a characteristic of the local practices in various countries. For instance, 20 bars per metre are quite common in France, whereas in the United States, slabs are frequently provided with 5 bars only, which are then of larger section, frequently square in section.

REINFORCED CONCRETE BRIDGES

7. Argument in Favour of n = 12

For normal working stresses n approximates 10. The ratio $n = \dfrac{E_s}{E_c}$ does not, however, remain constant, but varies, owing to the fact that the strain in the concrete increases at a higher rate than the stress. Under very small loads n is about 8, but at the ultimate stage, before breaking, it attains 22. This is the reason why the mean value, $n = 15$, is often adopted in the calculations. The ratio 12 is here introduced instead of the usual value 15, in order to impress upon the mind of the student the fact that all such assumptions are a matter of convention.

8. Footpath

A 600-kg/metre² distributed load is here assumed to represent the weight of the crowd. The slab is considered to be a cantilever, fixed in a point distant 10 cm (4 inches) from the outer face of the girder (see Fig. 111). Accordingly, the live moment is:

$$M_e = \frac{0.92^2}{2} \, 600 = 202 \text{ kg} \times \text{metres}.$$

Assuming a depth of slab of 12 cm and adopting a 2-cm layer of asphalt paving, the dead weight per square metre of surface is: $p_d = 0.12 \times 2{,}400 + 2 \times 30 = 288 + 60 = 348$ kg/metre², or in round numbers 350 kg/metre². Hence

$$M_d = \frac{0.92^2}{2} \, 350 = 148 \text{ kg} \times \text{metres}.$$

The total moment is therefore equal to $202 + 148 = 350$ kg × metres.

Using the same stress limits as before and the numerical coefficients from the tables A and C on pages 95 and 96 we find:

$d = 0.448 \sqrt{M} = 8.4$ cm, taken as 9 cm (total depth $= 9 + 3 = 12$ cm).

$a_s = 0.277 \sqrt{M} = 5.18$ cm², taken as 8 bars $\tfrac{3}{8}$ inch diameter per metre giving an area of $8 \times 0.71 = 5.68$ cm² (see table, p. 98).

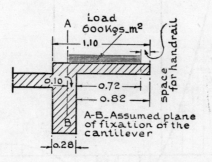

FIG. 111.

The longitudinal reinforcement will consist of 8 bars $\tfrac{1}{4}$ inch diameter per metre. This light reinforcement is intended to meet temperature stresses only, for there are no concentrated loads on the footpaths (note that the secondary reinforcement which we have placed in the main slab was chiefly intended to distribute such concentrated loads over a larger number of bars of the main reinforcement).

SECTION III. DESIGN OF CROSS-GIRDERS

1. Introductory

The cross-girder in this type of bridge forms a continuous beam resting on three or four supports (depending on the number of the main girders); but these supports are not as perfect as they are assumed to be in the common theory of continuous beams. The reasons are: (1) according to usual assumptions the reactions are concentrated at points, whereas in nature the cross-girder rests on the full width of the main girder, which causes the reaction to be distributed, and thus reduces the negative moment over the bearing; (2) the main girders are subject to elastic deflections, so that the theoretical bearings of the cross-girders are not entirely rigid but must be considered as being slightly elastic; this also reduces the effect of the negative moments.

For both these reasons, negative moments are generally less important in the design of reinforced concrete cross-girders than positive moments, and accordingly, we may allow higher values for the permissible stresses at the supports, than in the middle of the spans.

On the other hand, the bearings of the cross-girders are not hinged; consequently, the effect of the live loads in the adjacent spans upon the moment in the span considered, is taken into account only if it augments the stress, but not otherwise.

2. Dead-load Moments

In a continuous beam over four bearings (three equal spans, L metres long each) the dead-load moments are as follows:

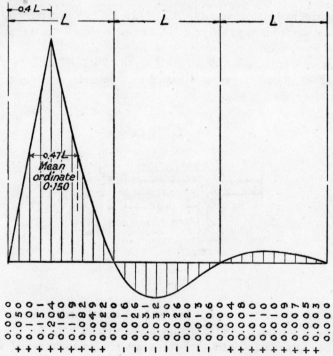

FIG. 112.—Influence line for continuous beams over four supports (3 equal spans). Positive bending moment at four-tenths ($0.4L$) of the span from end bearings.

REINFORCED CONCRETE BRIDGES

(1) Maximum positive moment in the lateral spans, at $0.4L$ from the end bearings:

$$(+M_d)_{max} = 0.64 \frac{pL^2}{8} = \frac{pL^2}{12}.$$

(2) Positive moment for intermediate span ($0.5L$ from the intermediate bearings):

$$(+M_d) = 0.2 \frac{pL^2}{8} = \frac{pL^2}{40}.$$

(3) Maximum negative moment over the intermediate bearings:

$$(-M_d)_{max} = 0.8 \frac{pL^2}{8} = \frac{pL^2}{10}.$$

3. Live-load Moments

The live-load moments will be found by means of the influence lines shown in Figs. 112 and 113.

The first curve gives the maximum positive moment at a distance $0.4L$ from the end bearing (same critical point as for the maximum dead moment).

The second curve gives the maximum negative moment over the intermediate bearing.

The author finds that a simplified formula may be used, instead of the influence line, for calculating the maximum positive live-load moment. In fact, the influence line for the case considered (see Fig. 112) forms nearly a triangle, the height of which is about $0.2L$. Consider a load P (see Fig. 114) distributed over a width $(n+e)$. The moment will then be equal to

$$M = \frac{P}{n+e} \text{ (area } abcds) \times L.$$

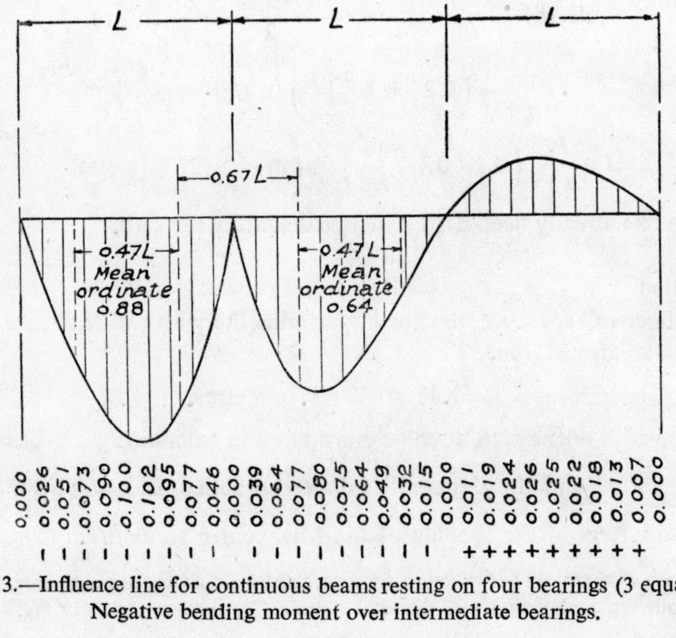

Fig. 113.—Influence line for continuous beams resting on four bearings (3 equal spans). Negative bending moment over intermediate bearings.

We may assume, in order to be on the safe side, that the lines ao and of are both straight (see Fig. 114).

Then
$$\frac{at}{ak}=\frac{bs}{of} \quad \text{and} \quad at=\frac{bs}{of}ak.$$

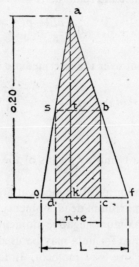

Fig. 114.

Consequently
$$bc=ak-at=ak-ak\frac{bs}{of}=0.2-0.2\frac{n+e}{L}.$$

The area
$$abcds=\frac{ak+bc}{2}dc=\frac{0.2+0.2-0.2\frac{n+e}{L}}{2}(n+e)$$
$$=\left(0.2-0.1\frac{n+e}{L}\right)(n+e).$$

Finally
$$M=\frac{PL}{n+e}\left(0.2-0.1\frac{n+e}{L}\right)(n+e)=\frac{P}{10}[2L-(n+e)].$$

This formula has already been used here in calculating the slab.

4. Design of Section

The total "effective" span, i.e. the clear span plus the width of the main girder, must be included in this calculation. Hence
$$L=1.48+0.28=1.76 \text{ metres.}$$

The effective width of the slab, b, to be considered in calculating a T-beam (according to the French specification and the frequently adopted practice) is: $b=\frac{L}{3}$; also, b should not be more than three-quarters of the spacing of the ribs, centre to centre.[1] Consequently in our

[1] In the case of a Γ-beam, that is to say, a rib crowned by a slab on one side only, the three-quarters criterion will be applied in conjunction with the total width of the slab, as shown in Fig. 111.

case, $b = \dfrac{1.76}{3}$ metre $= 58.67 \sim 59$ cm, which is less than $\tfrac{3}{4} \times 1.79 = 1.34$ metres.

In calculating the structural weight for the cross-girder, we may take the depth of the girder to be 32 cm, that is, about 13 inches (as an average for many bridges of this type). Accordingly, the dead load per metre length of the cross-girder is:

1. Weight of slab 1.79×570 $= 1{,}020$ kg/metre
2. Weight of rib $0.16 \times (0.32 - 0.16)\, 2{,}400 =$ 62 ,, ,,

                                                                      Total $1{,}082$,, ,,

taken as 1,080 kg/metre.

The corresponding bending moments, which we require for design purposes, are then found as follows:

$$\text{Maximum positive dead moment } (+M_d)_{max} = \frac{1{,}080 \times 176^2}{12} = 279 \text{ kg} \times \text{metres.}$$

$$\text{Maximum negative dead moment } (-M_d)_{max} = \frac{1{,}080 \times 176^2}{10} = 335 \text{ kg} \times \text{metres.}$$

To calculate the live-load moment, we first find the width upon which the weight of the wheel is distributed.

The value of e (see Fig. 115) to be introduced in this calculation is assumed equal to the effective depth of the girder, or about 6 cm less than its total depth. In fact, as shown in the same figure, the subtangent angle of the line of distribution is assumed to be 30°, which explains the figure quoted. Accordingly, the width over which the load is distributed, is

         Width of wheel $= 0.50$ metre
         Depth of girder $= 0.26$,, $= (0.32 - 0.06)$ metre
         Depth of pavement $= 0.06$,,

             Total 0.82 ,,   , which represents $\dfrac{0.8}{1.76} = 0.466$

of the span, and is taken as $0.47L$.

It will be noted that the branches of the first influence line shown in Fig. 112 for the span

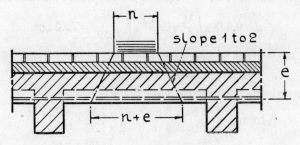

Fig. 115.

considered, are very nearly straight; consequently, it is easy to find the mean ordinate corresponding to $0.47L$.

As seen from the influence diagram in Fig. 112, this mean ordinate is in our case equal to 0.15.

Consequently *the maximum positive live moment* is:

$$(+M_e)_{max} = 6{,}000 \times 1.76 \times 0.15 = 1{,}585 \text{ kg} \times \text{metres.}$$

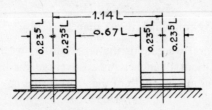

Fig. 116.

We neglect here the effect of the second wheel, because it falls in the middle span, and therefore tends to reduce the value of the moment due to the first wheel.

Note that the simplified formula would have given:

$$M = \frac{P}{10}[2L-(n+e)] = \frac{6,000}{10}(2 \times 1.76 - 0.82) = 1,620 \text{ kg} \times \text{metres},$$

which is very near to the precise figure calculated from the diagram (1,585 kg × metres).

For the maximum negative moment we must consider both wheels, because the effect of the second wheel increases the moment due to the first wheel.

The distance between the wheels is 2.00 metres. In units of span this represents $\frac{2.00}{1.76} = 1.137L$ or approximately $1.14L$. The diagram representing the two wheels is, therefore, as shown in Fig. 116.

On comparing this sketch with the influence diagram we find that the mean ordinate for the left-hand wheel is 0.083, and the mean ordinate for the right-hand wheel is 0.046. Thus *the maximum negative live moment* is:

$$(-M_e)_{max} = 6,000 \times 0.083 \times 1.76 + 6,000 \times 0.064 \times 1.76 = 1,552 \text{ kg} \times \text{metres}.$$

Recapitulating:

Total maximum positive moment $= M_e + M_d = 1,585 + 279 = 1,864$ kg × metres.
,,   ,,   negative   ,, $= M_e + M_d = 1,552 + 335 = 1,887$ kg × metres.

Design for Positive Moment. Taking the same permissible stresses as for the slab ($f_c = 35$ kg/cm², $f_s = 900$ kg/cm² and $n = 12$), the required effective depth is found to be:

$$d = 0.448\sqrt{\frac{M}{b}} = 0.448\sqrt{\frac{1,864}{0.59}} = 25.2 \text{ cm}$$

in which the coefficient 0.448 is taken, as before, from the table *A*, p. 95. The area of steel is:

$$a_s = 0.277\sqrt{Mb} = 0.277\sqrt{1,864 \times 0.59} = 9.18 \text{ cm}^2$$

the figure 0.277 being abstracted from the table *C* for coefficients, p. 96.

Adopted $\begin{cases} \text{2 bars } \frac{3}{4} \text{ inch} = 2 \times 2.86 = 5.70 \text{ cm}^2. \\ \text{2 bars } \frac{5}{8} \text{ inch} = 2 \times 1.98 = 3.96 \text{ ,,} \end{cases}$
Total 9.66 ,,

The depth of the girder is 32 cm (see section in Fig. 117).

Design for Negative Moment. Since in this particular case the lower part of the beam is compressed, whilst the upper part is extended, *b* cannot be taken wider than the rib = 0.16 metre. Assuming the permissible stresses $f_c = 40$ kg/cm², $f_s = 1,000$ kg/cm² and $n = 12$, we, therefore, have:

$$d = 0.416\sqrt{\frac{M}{b}} = 0.416\sqrt{\frac{1,887}{0.16}} = 45.2 \text{ cm}.$$

$$a_s = 0.27\sqrt{Mb} = 0.270\sqrt{1,887 \times 0.16} = 4.8 \text{ cm}^2.$$

REINFORCED CONCRETE BRIDGES

Note that the permissible stress limits embodied in this calculation are higher than for the positive moment because the bearings in this case do not realize the theoretical assumptions, and this reduces the negative moment.

Over the width $b=0.59$ metre the upper reinforcement of the slab is $0.59 \times 7.83 = 4.62$ cm². This steel would have been almost sufficient to carry by itself the calculated stress, but it is safer in this case to include in the calculation those only of the bars which are placed *directly* over the rib. Consequently, we will use 4 bars of $\frac{1}{2}$ inch diameter, giving an area of $4 \times 1.27 = 5.08$ cm² (see table, p. 98).

Two of the $\frac{1}{2}$-inch bars will then be straight and will extend over the entire length of the girder, while the two other ones will be bent (see Fig. 117).

The depth of the cross-girder, where it meets the main girder, is 50 cm (calculated theoretical depth: 45.2 cm, see above).

5. Checking the Stresses

Section in Middle of Span. Area of steel in tension: $a_s = 9.66$ cm².

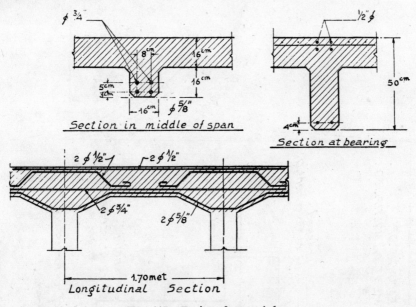

FIG. 117.—Design of cross-girder.

Distance from the under-surface of the rib to the centroid of reinforcement
$$= \frac{4 \times 3.96 + 9 \times 5.70}{9.66} = 6.96 \sim 7 \text{ cm.}$$
It follows that the distance from this centroid to the upper surface is $32 - 7 = 25$ cm. The proportion of steel multiplied by n is, therefore, equal to:
$$np = \frac{a_s \times n}{h \times b} = \frac{9.66 \times 12}{25 \times 59} = 0.0786.$$

Compressed reinforcement = 2 bars $\frac{1}{2}$ inch and 2 bars $\frac{3}{8}$ inch which gives 3.96 cm² (see table, p. 98).

Note: 2 bars $\frac{1}{2}$ inch diameter are placed directly above the extended bars and 2 bars $\frac{3}{8}$ inch diameter belong to the upper transverse reinforcement of the slab (see Fig. 117).

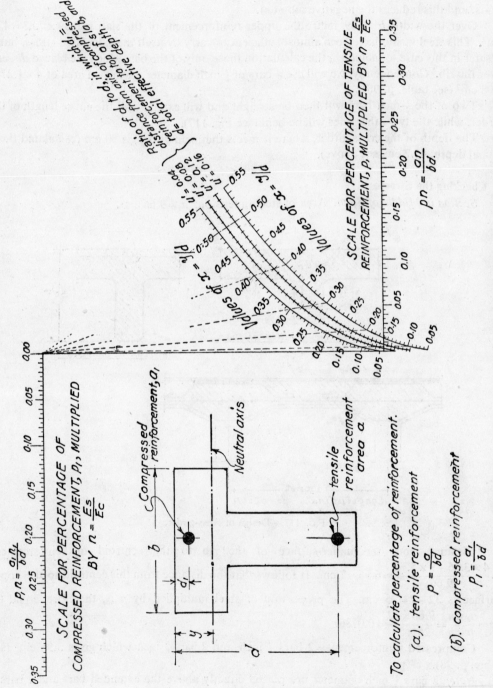

Fig. 118.—Alignment chart for finding the neutral axis in T-beams and slabs. Case when the neutral axis falls in the slab.

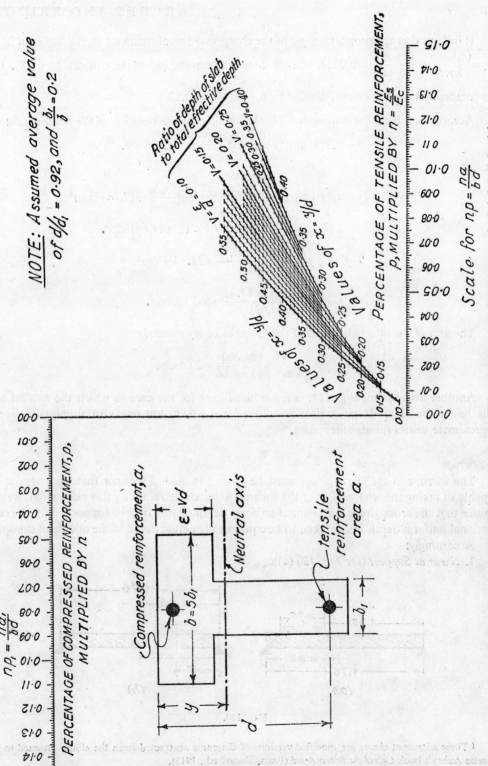

Fig. 119.—Alignment chart for finding the neutral axis in T-beams. Case when the neutral axis is below the slab.

It follows that the proportion, p_1, of the compressed reinforcement multiplied by n is, $np_1 = \dfrac{(a_s)_c \times n}{h \times b} = \dfrac{3.96 \times 12}{25 \times 59} = 0.0324$, and the distance from the top of the concrete to the C.G. of the compressed reinforcement, divided by h, is $u_1 = \dfrac{3}{25} = 0.12$.

Accordingly, from the nomogram Fig. 118 we get for the position of the neutral axis:

$$x = 0.311 \text{ and } y = xh = 0.311 \times 25 = 7.78 \text{ cm}.$$

Hence:

$$I = \frac{by^3}{3} + (y - u_1 h)^2 (a_s)_c n + (h-y)^2 a_s n = \frac{7.78^3 \times 59}{3} + (7.78 - 3.0)^2$$

$$\times 3.96 \times 12 + (25.0 - 7.78)^2 \times 9.66 \times 12 = 44{,}641 \text{ cm}^4.$$

$$f_c = \frac{M}{I} y = \frac{186{,}400 \times 7.78}{44{,}641} = 32.5 \text{ kg/cm}^2.$$

$$f_s = 12 \frac{186{,}400 \times 17.22}{44{,}641} = 863 \text{ kg/cm}^2.$$

The arm of the internal couple of the moment of resistance is:

$$j = \frac{M}{f_s \times a_s} = \frac{186{,}400}{863 \times 9.66} = 22.4 \text{ cm}.$$

Another nomogram, Fig. 119, is reproduced here for the case in which the neutral axis falls below the slab.[1] It gives precise values for $u_1 = 0.08$, but can also supply sufficiently approximate values for all other cases.

6. Stirrups

The stirrups in the cross-girder must be spaced in such a manner that they should be capable of taking the whole effect of the shearing stresses. To simplify this calculation, we can assume that the cross-girder is a beam of uniform depth resting on two supports only; the conventional uniform depth being taken to be equal to the actual depth in the middle of the span.

Accordingly:

1. *Shear at Support (see Fig. 120 (a))*:

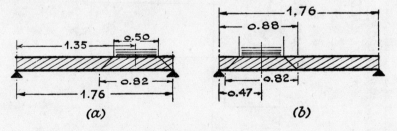

Fig. 120.

[1] These alignment charts are modified versions of diagrams abstracted from the album annexed to M. Charles Aubry's book *Calcul du Béton Armé* (Paris, Dunod ed., 1913).

A. Dead-load shear $= \dfrac{1.76 \times 1{,}080}{2} = 960$ kg.

B. Live-load shear $= \dfrac{6{,}000 \times 1.35}{1.76} = 4{,}600$ kg.

Total $\overline{5{,}460}$ kg.

2. *Shear in the Middle of the Span (see Fig. 120 (b)):*
   A. Dead-load shear $= 0$
   B. Live-load shear $= \dfrac{6{,}000 \times 0.47}{1.76} = 1{,}600$ kg.

Accordingly, the shearing stress is:

(a) At support $\quad \dfrac{5{,}460}{16 \times 22.4} = 15.2$ kg/cm²*

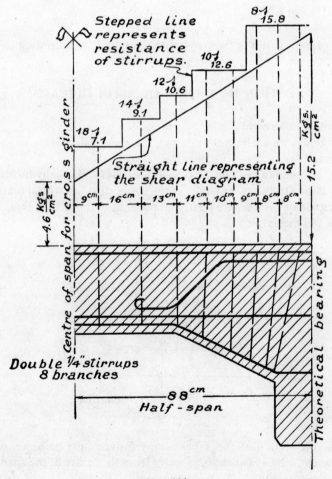

Fig. 121.

* This stress intensity is somewhat too high, but is admissible in this case, because the true depth of the rib in this spot is much greater than assumed in the calculation.

(b) In the middle $\dfrac{1,600}{16 \times 22.4} = 4.6 \text{ kg/cm}^2$

in which 16 cm is the width of the rib and 22.4 cm is the arm of the internal couple of the moment of resistance.

The stirrups will be made of bars 1/4 inch diameter, consisting of 8 branches each. Allowing 600 kg/cm² as the permissible stress for shear, the resistance of each stirrup will be (see table, p. 98):

$$8 \times 0.317 \times 800 = 2,040 \text{ kg.}$$

Assuming a spacing of stirrups 18, 12, 10 and 8 cm, the resistance of the stirrups will be equivalent to the following stresses:

$\dfrac{2,040}{18 \times 16} = 7.1 \text{ kg/cm}^2$ $\qquad$ $\dfrac{2,040}{10 \times 16} = 12.6 \text{ kg/cm}^2$

$\dfrac{2,040}{12 \times 16} = 10.6$,, ,, $\qquad$ $\dfrac{2,040}{8 \times 16} = 15.8$,, ,,

The spacing of the stirrups is then computed graphically, according to these figures, as shown in Fig. 121.

SECTION IV. DESIGN OF MAIN GIRDERS

1. Order of Procedure in General

In the type of bridge we are here considering, the main girder forms a T-beam, simply supported at either end.

In the middle of the span the neutral axis of this beam will usually be found to fall within the rib, below the slab. On the other hand, at the supports it lies within the slab (this follows from the fact that some of the bars are bent, and the area of the steel in tension is therefore reduced towards the abutments).

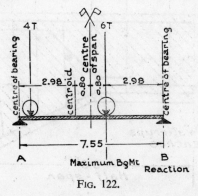

Fig. 122.

The design of the section of the T-beam may conveniently be begun with the point of maximum live moment, which is about 0.80 metre from the centre of the span (with the standard 20-ton lorry shown in Fig. 122). Note that, although the points of maximum moments due, respectively, to live and dead loads do not coincide, the usual practice is to consider in the calculation the sum of these moments; this is, therefore, slightly greater than the theoretical maximum.

REINFORCED CONCRETE BRIDGES

As the maximum moment occurs near the middle of the span, the main girder is designed on the assumption that the neutral axis falls below the slab (as shown in Fig. 123).

2. Analytical Formulae

The precise formulae, applicable to the present case, are:

(1) *To find the neutral axis we should use the "equation of moments":*

$$\frac{y^2 b_1}{2} + D(b-b_1)\left(y-\frac{D}{2}\right) + na_1(y-q) - na(d-y) = 0. \tag{1}$$

in which y is the required distance from the top of the concrete to the neutral axis,
   b is the effective width of the slab introduced in the calculation,
   b_1 is the width of the rib,
   D is the depth of the slab,
   a is the area of the tensile reinforcement,
   a_1 is the area of the compressive reinforcement,
   d is the distance from the top of the slab to the C.G. of the tensile reinforcement,
   q is the distance from the top of the slab to the C.G. of the compressive reinforcement.

This equation will yield the value of y.

(2) *To find the moment of inertia (second moment):*

$$\frac{y^3 b}{3} - \frac{(y-D)^3 (b-b_1)}{3} + na_1(y-q)^2 + na(d-y)^2 = I \tag{2}$$

(3) *To find the stresses f_c and f_s in the concrete and in the steel:*

$$f_c = \frac{My}{I}.$$

$$f_s = n\frac{M(d-y)}{I}.$$

These formulae can be easily developed, but they involve rather laborious calculations when applied in practice. For example, let us find the value y from equation (1).

This equation may, then, be transcribed as follows:

$$y^2 + \frac{2y}{b_1}[D(b-b_1) + n(a_1+a)] - \frac{D^2}{b_1}(b-b_1) - \frac{2na_1 q}{b_1} - \frac{2nad}{b_1} = 0.$$

Solving:

$$y = -\frac{D(b-b_1) + n(a_1+a)}{b_1} + \sqrt{\left(\frac{D(b-b_1) + n(a+a_1)}{b_1}\right)^2 + \frac{2}{b_1}\left[\frac{D^2}{2}(b-b_1) + n(a_1 q + ad)\right]}.$$

It is clear that the application of this formula in practice, is by no means an easy operation.

3. Graphical Calculation—Method Recommended

In order to simplify these calculations a graphical method has been successfully used, which is described below.

The sectional area of the compressed portion of the beam is divided into a series of parallel strips, as shown in Fig. 123. The area of each strip is then assimilated to an equivalent force applied in its centroid, and for this set of equivalent forces, a funicular polygon is computed in the usual way. Note that the area of the compressed reinforcement multiplied by $(n-1)$ is visualized as one of the forces thus described, and is therefore included in the computation of the polygon.

The area of the tensile reinforcement, multiplied by n, is considered, in drawing the stress diagram, as a negative force, viz. the direction of this force is opposite to the direction of the forces which are taken to represent the compressed area. A new polygon is then drawn for the tensile reinforcement, as shown in the drawing.

It is known from elementary statics, that every ordinate ab (see drawing) in the compressed part of the polygon represents to a certain scale the first moment of the corresponding area.

On the other hand, every ordinate a_1b_1 of the extended polygon (line AC) represents, to the same scale, the first moment of the reduced area of the tensile reinforcement.

Consequently, at the point C, in which the lines AC and BC meet, the first moments of the compression area and of the reduced tension area are both equal.

It follows that the neutral axis is determined by the intersection of the polygons AC and BC, in Fig. 123.

The moment of inertia I (second moment) may then be directly obtained from this drawing by the planimeter, as twice the combined area ABC, and thereafter, the stresses may be found from

$$f_c = \frac{My}{I} \text{ and } f_s = \frac{M(d-y)}{I} n.$$

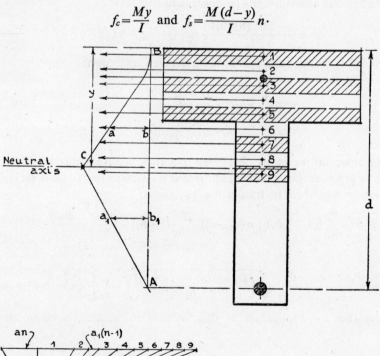

Fig. 123.

REINFORCED CONCRETE BRIDGES

4. Approximate Analytical Solution, Convenient for Preliminary Analysis

This method may be conveniently used for the final calculation of the T-section; but, in order to select *approximately* the main elements of that section, some other, simplified solution must be used as a guide for the preliminary design.

When the neutral axis falls close to the slab, it may be considered, in the first approximation, that the total depth of the slab is equal to the distance between this axis and the top of the concrete. We may then employ the usual equations applicable to slabs:

$$d = \theta \sqrt{\frac{M}{b}}$$

$$a = \gamma \sqrt{M \times b}.$$

in which θ and γ are the coefficients listed in the tables on pages 95 to 97.

The value d found from this equation must, in such cases, be slightly increased, because the conventional application of the slab formula to a T-beam reduces the safety of the calculation.

On the other hand, the effect of this convention is very small and is to a certain extent compensated by neglecting the compressed reinforcement; at any rate, verification of the designed section by means of the precise method is strongly recommended.

Generally speaking, it appears from a large number of such calculations that the error may be from 1 per cent to 20 per cent.

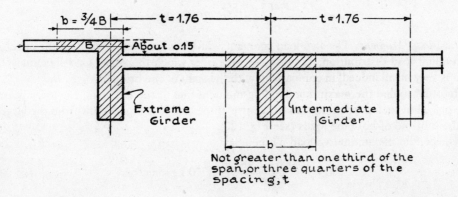

Fig. 124.

5. Calculation of Bending Moments

As intimated earlier, the width, b, of the part of the slab (see Fig. 124) which is considered as acting in conjunction with the rib, is either equal to one-third of the span or to three-quarters of the centre-to-centre spacing of the girders, whichever of these two values is the smaller. For the main girders, the second condition is usually the significant criterion. Accordingly, in our case,

$$b = 0.75 \times 1.76 = 1.32 \text{ metres.}$$

Note that for the extreme girder, the adopted value of b must not exceed three-quarters of the width of the footpath, as shown in the drawing.

Dead-weight Moment. The standard formula applicable to the case is:

$$M_d = \frac{p_d L^2}{8}.$$

In calculating the dead-load moment, the weight of each cross-girder is assumed, for the sake of simplicity, to be uniformly distributed over a distance equal to the spacing of the cross-girders. In the same order of ideas, the depth of the main girder may be taken from one-seventh to one-fourteenth of its clear span. The first value is for spans of about 4 to 5 metres; the second value applies to spans of about 13 metres. Generally speaking, it must be remembered that in calculating the bending moment for the main girder, a small difference in the depth of the rib is of *little importance,* for it affects the weight distribution only very slightly.

In the calculation under review, for the estimate of the structural weight, the total depth of the girder was approximately taken to be 0.73 metre and its width 0.28 metre.

Accordingly, the dead weight, p_d, per metre length of the main girder was:

(1) Weight of slab $\quad 1.76 \times 570 = 1{,}003$ kg/metre
(2) Weight of cross-girder $\quad \dfrac{0.16 \times 1.48 \times 2{,}400}{1.79} \cdot \left(\dfrac{0.32 + 0.50}{2} - 0.16\right) = 79$,,
(3) Weight of main girder $\quad (0.73 - 0.16) \times 0.28 \times 2{,}400 = 383$,,
$\hspace{5cm}\text{Total} = 1{,}465$,,

Consequently

$$M_d = \frac{p_d L^2}{8} = \frac{1{,}465 \times 7.55^2}{8} = 10{,}430 \text{ kg} \times \text{metres}.[1]$$

Live-load Moment. The live-load moment will be calculated neglecting the effect of the longitudinal load distribution, because the area over which the wheel load is actually distributed is very small indeed, in comparison with the span of the bridge.

In order to find the maximum bending moment due to the live load, the lorry is assumed to be placed in such a position that the centre of the bridge is equidistant from the larger load and the centroid of both the loads (see Fig. 122).

Hence, the maximum live-load moment is:

$$M_e = \frac{10{,}000 \times 2.98^2}{7.55} = 11{,}770 \text{ kg} \times \text{metres}$$

in which 2.98 metres is the distance from the main wheel to the bearing while 7.55 is the effective span.

The total moment is then equal to:

$$M_{total} = M_d + M_e = 10{,}430 + 11{,}770 = 22{,}200 \text{ kg} \times \text{metres}.$$

6. Preliminary Design, by Means of Approximate Formula

In designing the main girder we shall assume $f_c = 35$, $f_s = 1{,}100$, which are larger than before because the effect of impact is in this case less, and $n = 12$. In regard to safety, the permissible limit for concrete could have been still increased, but this would have involved an excessive number of bars. The selected relation between permissible stresses gives a sound,

---

[1] In this calculation, 7.55 metres is the effective, theoretical span, corresponding to a clear span of 7.00 metres.

REINFORCED CONCRETE BRIDGES

economical proportion of steel to concrete, remembering that no impact load is included explicitly in this calculation.

The "slab" formula gives, in this case, the following approximate values:

$$d = 0.477 \sqrt{\frac{M}{b}} = 0.477 \sqrt{\frac{22,200}{1.32}} = 61.9 \text{ cm}$$

$$a = 0.21 \sqrt{M \times b} = 35.9 \text{ cm}^2.$$

Note that the coefficients 0.477 and 0.21 appearing in these formulae are taken from the tables A and C on pages 95 and 96.

We shall assume: $d = 63$ cm and $a = 9$ bars 7/8 inch dia $= 9 \times 3.88 = 34.9$ cm^2.

The compressed reinforcement is formed of 3 bars 3/4 inch dia $= 8.55$ cm^2.

7. Verification by Graphical Method

The check will be made by means of the graphical method (see Fig. 125).

The area of the compressed concrete is divided into strips 2 cm deep each. The area of each strip in the slab (strips 1 to 8) is therefore $2 \times 132 = 264$ cm^2. In the rib, the area of each strip is $2 \times 28 = 56$ cm^2.

The additional area, representing the compressed reinforcement, is $(12-1) \times 8.55 = 94$ cm^2. This area is added to the area of strip 3, because the compressed reinforcement is placed at 5 cm from the top of the slab, which corresponds to the centre of the strip referred to.

In drawing the stress diagram for the funicular polygon, it is preferable to make the summation of areas arithmetically, and then to plot, in the diagram, the accumulative areas thus obtained. The calculation is shown in the following table:

NN of strip	Area, cm^2	Accumulative area, cm^2
1	264	264
2	264	528
3	264 + 94	886
4	264	1,150
5	264	1,414
6	264	1,678
7	264	1,942
8	264	2,206
9	56	2,262
10	56	2,318

The figures representing the accumulative area are plotted on the right-hand side of the drawing in Fig. 126 to the scale 1 cm to 200 cm^2.

For convenience of the computation of the funicular polygon, the polar distance must be selected as about one-half to one-third of the vector representing the total accumulative area. Accordingly, we take as polar distance

$$H = 1,000 \text{ cm}^2 \text{ or, to scale, 5 cm.}$$

The funicular polygon is then constructed in the left-hand bottom corner of the same drawing.

The scale for lengths is 1 to 2.5.

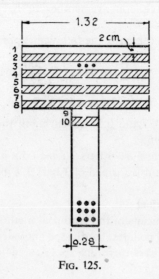

Fig. 125.

The tensile area may now be calculated in the same way as the compressed area:

Total number of bars	Area × n		Accumulative area × n
3	3 × 3.88 × 12	=	140
6	6 × 3.88 × 12	=	279
8	8 × 3.88 × 12	=	372
9	9 × 3.88 × 12	=	419

Usually the tension area of the steel is much smaller than the compressed area of concrete. In drawing the funicular polygons it is therefore advisable to select for the tension zone a scale of areas twice as large as that of the compressed zone; namely, in our case, 1 cm to 100 cm². The polar distance for the tension polygon will then be twice as great as for the compression diagram, as shown in the drawing.

The intersection points of the two polygons supply the positions of the neutral axes for different numbers of bars, because at these points the first moments, of steel and concrete, respectively, are equal. The values y, as found from the drawing, are then as follows (see left-hand bottom corner of diagram):

3 bars 7/8 inch	$y = 10.5$ cm	$d - y = 52.5$ cm
6 bars 7/8 inch	$y = 14.1$,,	$d - y = 48.9$,,
8 bars 7/8 inch	$y = 15.8$,,	$d - y = 47.2$,,
9 bars 7/8 inch	$y = 16.5$,,	$d - y = 46.5$,,

The areas of the polygons corresponding to 3 bars, 6 bars, 8 bars and 9 bars respectively are then measured by means of the planimeter and the following figures are thus obtained:

3 bars	40.6 cm²
6 bars	69.0 ,,
8 bars	83.4 ,,
9 bars	89.2 ,,

REINFORCED CONCRETE BRIDGES

8. Question of Scales

Each of these figures represents, to a certain scale, one-half of the corresponding second moment. It remains to ascertain this scale. This is a point which, in the drawing office and schoolroom, frequently leads to errors. We shall, therefore, consider it here in detail.

Suppose that ST (see Fig. 127) represents an area A cm², to a scale 1 cm to N cm². Then, the length ST in the drawing is equal to A/N cm $= t$ cm. Let the polar distance be F cm² or $\frac{F}{N} = 5$ cm in the drawing.

Consider the first moment and the moment of inertia of the area A about a point X, which lies at a distance Y from that area (see Fig. 127). Suppose the left-hand sketch to be drawn to a scale 1 cm to M cm and let us draw the funicular polygon, which in this case is reduced to the straight line PL, parallel to the line QS. The distance PX is equal to $\frac{Y}{M}$; consequently,

$$\frac{LX}{PX} = \frac{ST}{QT}$$

$$LX = \frac{ST \times PX}{QT} = \frac{\frac{A}{N} \times \frac{Y}{M}}{\frac{F}{N}} = \frac{AY}{FM} = \frac{\text{First moment}}{MF}.$$

Hence the first moment $=$ (Ordinate LX) $\times MF$.

Transcribing this equation in words: the first moment is equal to the ordinate of the funicular polygon, multiplied by the scale of lengths and by the area represented by the polar distance. Each centimetre of the ordinate corresponds to MF centimetres-cube of the first moment.

Accordingly, in our case, i.e. in Fig. 126, the scale of the ordinates is 1 cm to $MF = 2.5 \times 1{,}000 = 2{,}500$ cm³.

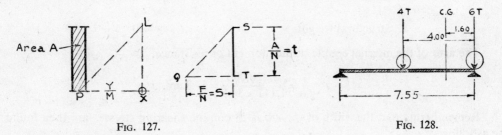

Fig. 127.   Fig. 128.

Let us now consider the area of the triangle PLX. It is equal to

$$\frac{1}{2} PX \times LX = \frac{1}{2} \frac{Y}{M} \times \frac{AY}{FM} = \frac{AY^2}{2} \frac{1}{FM^2} = \frac{\text{Moment of inertia}}{2}.$$

It follows, that the area of the polygon represents half the moment of inertia, to the scale FM^2. In our case $FM^2 = 1{,}000 \times 2.5 \times 2.5 = 6{,}250$.

Consequently, in Fig. 126 the scale is 1 cm square to 6,250 cm⁴.

Thus, the moments of inertia, found from the diagram, are

(1) for 3 bars 7/8 inch $2 \times 6{,}250 \times 40.6$ cm² = 507,000 cm⁴
(2) for 6 bars 7/8 inch $2 \times 6{,}250 \times 69.0$,, = 862,000 ,,
(3) for 8 bars 7/8 inch $2 \times 6{,}250 \times 83.4$,, = 1,042,000 ,,
(4) for 9 bars 7/8 inch $2 \times 6{,}250 \times 89.2$,, = 1,115,000 ,,

This means that the stresses f_c and f_s at the middle point of the span are respectively equal to:

$$f_c = \frac{My}{I} = \frac{2{,}220{,}000 \times 16.5}{1{,}115{,}000} = 32.9 \text{ kg per sq cm.}$$

$$f_s = \frac{M(d-y)}{I} n = \frac{2{,}220{,}000 \times 46.5 \times 12}{1{,}115{,}000} = 1{,}111 \text{ kg per sq cm}$$

which is close enough to the values assumed in the preliminary calculation, i.e. $f_c = 35$ kg per sq cm and $f_s = 1{,}100$ kg per sq cm, and is therefore admissible.

9. Bent Bars and Spacing of Stirrups

The arrangement of the bent bars and the spacing of the stirrups will depend primarily on the shearing stresses and diagonal tension, but they must also be verified with reference to the diagram of bending moments.

Shear and Shearing Stress. (a) The maximum shear at the bearing is (see Fig. 128).

1. Live load $\dfrac{10{,}000\,(7.55 - 1.6)}{7.55} = 7{,}880$ kg

2. Structural weight $\dfrac{1{,}465 \times 7.55}{2} = 5{,}530$,,

                  Total   13,410 ,,

(b) In the middle of the span the shear is:

1. Live load $\dfrac{6{,}000}{2} = 3{,}000$ kg

2. Structural weight   = 0

The arm of the internal couple of the moment of resistance,

$$j = \frac{M}{f_s a} = \frac{2{,}220{,}000}{1{,}111 \times 34.9} = 57.3 \text{ cm.}$$

Remembering that the width of the rib is 28 cm, the shearing stresses are then found to be as follows:

(a) At the bearing $\dfrac{13{,}410}{28 \times 57.3} = 8.36$ kg × cm²

(b) In the middle $\dfrac{3{,}000}{28 \times 57.3} = 1.87$,, ,,

To design the arrangement of the diagonal bars the intensity of shearing stresses will be assumed equal to the diagonal tension stress.

The diagram of shear is shown in Fig. 129. It is seen from this diagram that the shearing stress exceeds the limit of 4 kg/cm² over the portion of the span extending from the support to a point located at about 1.25 metres from the middle of the girder.

For this portion all the shear must, therefore, be taken by the metal. We shall adopt a spacing of stirrups equal to 12 cm. Each stirrup will be formed of 6 branches of 1/4 inch diameter bars, thus giving a total sectional area of $6 \times 0.317 = 1.902$ cm². Allowing a stress of 800 kg/cm² for shear, each stirrup may, consequently, take $1.902 \times 800 = 1,521$ kg. Accordingly, the shearing stress taken by these stirrups is:

$$S_1 = \frac{1,521}{28 \times 12} = 4.5 \text{ kg/cm}^2.$$

This value is represented by the hatched area in the diagram, Fig. 129.

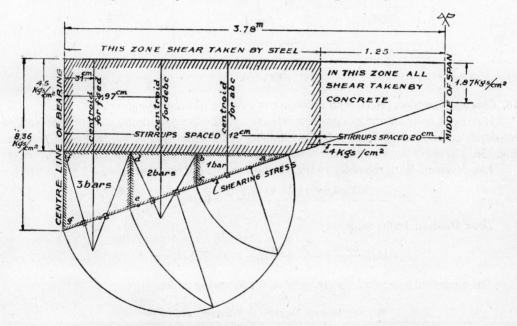

Fig. 129.—Diagram for the spacing of stirrups and arrangement of bent bars.

As seen from the drawing, the stirrups spaced in this manner take only part of the total shear. The rest must, consequently, be taken by the bent bars.

In the middle of the span we have 9 tension bars. Three of these, corresponding to the three upper bars, must be run right through to the bearing, in order to facilitate the erection of the stirrups, but the remaining six bars may be bent. They will be arranged in the following manner: first, one bar will be bent; then two bars, and thirdly, three bars, near the bearing.

The triangular shear diagram, which represents the stress taken by these bars, is, consequently, divided into: one-sixth, two-sixths and three-sixths parts. The usual graphical computation is employed. As seen from the drawing Fig. 130, this gives the following spacings: 163, 97 and 31 cm from the centre of the bearing. These distances determine the location of the points of intersection of the bent bars with the neutral axis. The apexes of the angles are located about 45 cm farther from the bearing (see Fig. 130); namely, at 208, 142 and 76 cm from the same point.

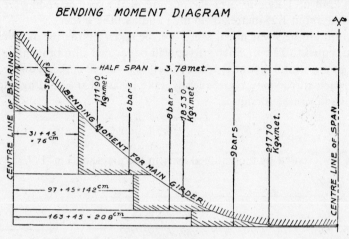

Fig. 130.

10. Checking Lengths of Bars, with Reference to Bending Moment Diagram

It remains to check the design with reference to the bending moment. The moments of resistance corresponding to the various numbers of bars are used in preparing the graph in Fig. 130. The latter also contains the bending moments calculated from the following formulae:

Live Moment. With reference to the notation in Fig. 131 we have

$$M_e = \frac{(6{,}000 + 4{,}000)\,(L - x - 1.6)\,x}{L} = 10{,}000\,\frac{L - x - 1.6}{L}\,x.$$

Dead Moment. In the same way

$$M_d = \frac{L p_d x}{2} - \frac{x^2 p_d}{2} = \frac{p_d}{2}(L - x)\,x = 735\,(L - x)\,x.$$

The numerical computations are listed in the following table:

Bending Moment Diagram in Kilograms and Metres

	At centre	At 0.80 metres from centre	At 1.80 metres from centre	At 2.80 metres from centre
x	3.78	2.98	1.98	0.98
$L - x$	3.78	4.58	5.58	6.58
$(L - x)x$	14.29	13.65	11.05	6.45
$M_d = \frac{(L-x)xP}{2}$	10,430	10,000	8,100	2,730
$L - x - 1.6$	—	2.98	3.98	4.98
$x(L - x - 1.6)$	—	8.88	7.88	4.88
$M_e = 10{,}000\,\frac{x(L - x - 1.6)}{L}$	11,340	11,770	10,430	6,460
Total M	21,770	21,770	18,530	11,190

These results are then plotted in Fig. 130, which contains, also, the diagram of the moments of resistance for different numbers of tensile bars, according to the proposed arrangement

REINFORCED CONCRETE BRIDGES

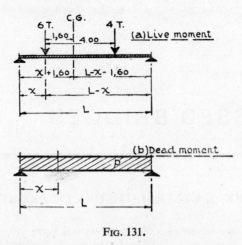

Fig. 131.

of bends, as found earlier from considerations concerning the shear diagram.

Comparing now the two diagrams in Fig. 130, the arrangement in question is found to be safe, also from the standpoint of the requirements of the bending moment diagram. Had this not been the case, a fresh trial would have been made.

Chapter Five
PRE-STRESSED BRIDGES

SECTION I. GENERAL THEORY OF PRE-STRESS

1. The Interdependence of the Advantages of Pre-stress and Span

As seen from the foregoing design example, the live load for a relatively short-span reinforced concrete T-beam bridge is rather more important than its structural weight. There is, therefore, no particular advantage in such works to depart from the orthodox type, in regard to both material and design.

As, however, the span increases even slightly, the structural weight becomes gradually more important, and there may then be good reasons to alter the usual design type; i.e. to use materials which, though more expensive (insofar as unit prices are concerned) allow nevertheless the raising materially of the permissible limits of stress, and thus reduce the structural weight of the bridge. An over-all economy may then be achieved because of a lighter structure, notwithstanding the fact that the materials employed in building it, i.e. high-strength steel and concrete, are more expensive than the current class employed in ordinary reinforced concrete designs.

The trouble, however, with these high-class materials is that, as steel is being subjected to higher and higher stresses, its deformations increase and cause the adjacent concrete to crack—in a manner objectionable for the safety of the work. This was one of the considerations which gave rise to the idea of compressing this concrete artificially, i.e. pre-stressing it, before (and partly, after) it had been put to work. It is, therefore, clear that pre-stressing is of particular advantage for long-span bridges.

On the other hand, apart from bridges, pre-stressing has now found its way into a large number of other engineering designs, such as circular tanks, shell-roofs, railway sleepers, etc., and the scope of its application goes on ever increasing, from day to day. It is possibly symptomatic of the interest devoted nowadays to the pre-stressing technique, that countless books are published and papers and articles on the subject are scattered about the pages of modern technical magazines and up-to-date issues of the proceedings of various engineering societies, giving stimulating ideas about new applications of the method.

2. Principle of Pre-stress

The principle of pre-stressing is, in fact, a rather obvious, and not entirely new idea. In order to understand it, let us imagine a bridge, carrying a number of loads a, b, c, etc., or a uniformly distributed loading, p. The bridge will deflect as shown by the dotted lines $\alpha\beta$ in

PRE-STRESSED BRIDGES

Fig. 132 (*A*), and the stress will then be tension in the lower part of the beam and compression in the upper part (see diagram *I–II* in the same figure). These are facts of common knowledge, which do not call for further explanation.

Consider, now, figure (*B*), appearing in the same drawing. Here we have inserted a cable, or a wire, in the lower portion of the girder, and by tightening it up—by means of hydraulic jacks or in any other convenient manner—we have created a compressive force N, which, in turn, has caused the bridge to camber, as shown by the dotted line δy in the same drawing. As seen from the diagram drawn in figure (*B*), we have, now, tension in the upper part of the beam and compression in its lower part, i.e. the stresses in (*B*) are opposite to those which we have observed in (*A*).

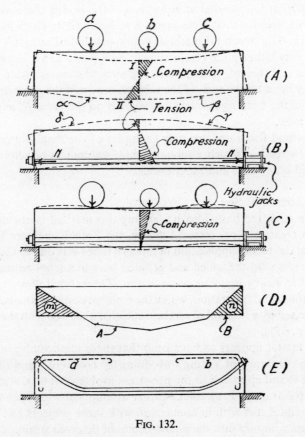

Fig. 132.

Hence, had the tightening-up operation—or, to use the more common term, the pre-stress—been applied before the vehicles were permitted to cross the bridge, there would have been compensation of stress after the bridge was opened to traffic, and if properly calculated and designed, the beam would have been subject to compressive stresses alone, as shown in the diagram in figure (*C*).[1] This, in a nutshell, is the argument of the pre-stress.

---

[1] In the case of permanent external loads, such as the structural weight, both loadings must, of course, be applied concurrently.

But the reader is warned that presented in this manner, the theory of the method is an oversimplified version of this rather involved technique, which, in practice, has to cope with numerous incident *sui generis* problems. For instance, the force N applied by means of the jacks will not remain constant in the future, but will depend on (a) the shrinkage and (b) the creep of the concrete, and this contingency is a serious consideration which must be taken into account in designing pre-stressed work.

On the other hand, attention may also be called to the fact that the bending moment produced by the force N is constant for all the length of the bridge (see diagram B in figure (D), in the same drawing), whereas for a simply supported beam the bending moment consequent on the effect of external forces, such as the loads a, b, c, tends to zero towards the bearings (see A in the same figure (D)). It follows that at either end of the bridge there will remain an unbalanced bending moment (see hatched areas m and n in figure (D)), which may be dangerous for the safety of the work. The structural methods employed to deal with this latter problem may be of two types: we may either place reinforcement bars (see a and b in figure (E)), at the top of the girder at either end thereof, so as to take the tension stresses produced by the cable; or alternatively, we may bend up the ends of the wire, as shown in the same drawing, and thus reduce the magnitude of the bending moment consequent on the force N, at the particular spots where it may be harmful.

It will be understood that these are only examples of the manifold problems incident to the technique of pre-stressing, which although simple enough in principle, is nevertheless, a matter for specialized knowledge and experience.

3. Basic Idea of Pre-stress Not New

As already intimated, the basic idea underlying this method is not new. In outlining its history[1] the modern creator of pre-stressed design, the French engineer Eugène Freyssinet,[2] considers that its first deliberate application in modern times was due to Considère, who in the early nineteen hundreds, used tensioned and grouted bars to anchor to rock masonry-blocks, which were exposed to heavy breaking-wave action. Thereafter followed, however, a disappointing period of failure and frustration, which the eminent designer attributes to the "intellectual oppression exercised by a handful of mathematicians, obsessed with their science and blind to reality" (*loc. cit.*, p. 331).

The fact of the matter appears to have been that these mathematicians held "an unquestionable belief in the constancy of Young's Modulus for concrete", and this belief stood actually in the way of a sound approach to the pre-stress problem. It was, also, the cause of dangerous, unforeseen strains in M. Freyssinet's earlier attempts at wide-span arches of pre-stressed type. But the troubles met with in connection with these projects had a useful result, for they prompted a deeper inquiry into the phenomenon of deferred strain.

By 1926–27, research carried out by Glanville in England and by Faber in America provided confirmation to M. Freyssinet's own results and afforded evidence reaffirming the possibility of maintaining high permanent stress in concrete. Thus it was that, in 1934, he was able to use his theory in connection with the consolidation of the maritime railway station at Le Havre—

---

[1] "Pre-stressed Concrete", Lecture delivered at the Institution of Civil Engineers, London, on Nov. 17, 1949. See *Journal* of the Institution, No. 4, 1949–50, p. 331.

[2] It will be observed that we use here the term "creator" and not "inventor", because this eminent engineer says (*loc. cit.*, p. 337): "The principles underlying pre-stressing are....a phenomenon of nature, and to speak of the 'invention of pre-stressing' is an absurdity."

the first daring application of the pre-stress technique to large-scale public works. Since that memorable date, it developed at a rapid rate both as regards theory and method, as well as application.

To this brief résumé of the modern evolution of the pre-stress principle we may possibly add that it was first used, though not in connection with concrete, at an earlier date than M. Freyssinet seems to think; in fact, there are good reasons to believe that it was employed by the mediaeval Swiss master-craftsmen responsible for timber bridges of surprisingly wide spans, dating back to this period. It was also known to the Slavonic artisans in Galicia, who worked in 1915 under the direction of Russian army engineers in restoring the bridges destroyed by the retreating Austrians.

4. Continuous and Discontinuous Structures

Another significant and enlightening point made out by M. Freyssinet, is the distinction he suggests between the pre-stress in "continuous" and in "discontinuous" structures.

The first group are wholly elastic structures, for the full range of stresses under which they work (or are designed to work). Within this range—and even beyond it, in the plastic domain—the stress-strain relation for such structures is fundamentally *continuous* and cannot be changed by pre-stress.

These are, for instance, riveted or welded steel work, laminated timber-work as used in aircraft construction, and so on. In such structures pre-stress is employed in exceptional cases only, chiefly in connection with permanent loading.

There is, however, another category of structures—Freyssinet's "discontinuous" type—in which an increase of the stress beyond a certain, well-defined point causes abrupt *discontinuities* in the stress-strain function, to wit: a masonry arch, or a reinforced concrete beam, or structures of the same class, i.e. behaving in the same manner under load.

Consider a masonry arch built of ashlar voussoirs, with joints having no material tensile strength. When loaded with a gradually increasing concentrated load, the deformations of such an arch will remain continuous, so long as the resultant will pass within the core, i.e. within the middle third of every section. But, as soon as this limit is exceeded, the rate of the increase of the deformation will suddenly rise, because part of the joint will then be open, and soon thereafter failure will take place, when the resultant approaches the border.

Failure will occur in this case independently (or almost so) of the crushing strength of the material, but characteristically, no permanent deformation will remain, should the rate of the increase of the load be reversed, within the period between the appearance of the first signs of weakness and the final failure. In this, the second type, the structure differs from the first, for with the continuous type, permanent effect is produced as soon as the elastic limit is exceeded, and the deformations show a tendency to increase at a more rapid rate than the stresses.

Another example of the discontinuous type is the reinforced concrete beam, for here again, the deformation is slow and gradual at the beginning, until plasticity takes control of the phenomenon and cracks start to appear in the extended portion of the material; the rate of the deformation is then observed to increase suddenly.

It is for this second type of structures that pre-stress is particularly efficacious, for by preventing tensile stresses, or keeping them below the tensile strength of the material, we control the stress-strain relation to our advantage and thus improve materially the resistance of the structure.[1]

[1] The foregoing is only a summary of M. Freyssinet's argument. The original will repay study.

It might be mentioned, incidentally, that M. Freyssinet's suggestion to distinguish between the so-called "continuous" and "discontinuous" structures, is a modern approach, which does not belong to the theory of pre-stress alone, but goes much further than that, for it contributes towards a better understanding of the argument of certain recent tendencies in connection with various problems of structural philosophy.

In fact, so long as we deal with continuous structures only, the ratio of the breaking stress to the working stress, though *not exactly* equal to the ratio of breaking load to actual load, is nevertheless an approximation thereto; but, when the scope of the inquiry is extended to include discontinuous structures also, it is found that the two ratios in question are *not even approximately* equal, for a slight increase in the load may frequently alter the nature itself of the stress-strain relation (as in the case of the arch examined earlier), and the danger of failure will then be suddenly aggravated.

5. Non-Elastic Behaviour of Materials

The significance of the term "not exactly" appearing in the foregoing paragraph (in italics), as contrasted with the term "not even approximately" (also in italics), deserves, perhaps, being explained more fully. In fact, taking reinforced concrete as an example, we shall find that until the conventional elastic limit is reached, the deformations of the steel reinforcement are almost proportional to the stresses and the stress-strain diagram for this material is, consequently, almost exactly straight. On the other hand, for concrete, the diagram is slightly curved from the start, and therefore strictly speaking, the rate of the increase of the deformations differs from that of the stresses, as early as the loads are first applied.

However, taking the tangent to the curve of deformation in the neighbourhood of its origin as the basic parameter, and considering it as determining the elastic modulus, we can delimit a *nearly-elastic* domain, within the boundaries of which the common theory of the resistance of materials will remain approximately true. This is the domain in which we usually operate, when designing reinforced concrete in practice. In these limits strains are nearly elastic, but not linearly so.

As, however, stresses increase, the effect of the curvature of the stress-strain diagram becomes more pronounced, and at the moment of failure, plastic deformation has as great an influence as (or even a greater influence than) the elastic strains.

In view of these latter facts, frequent suggestions have been put forward, in the recent past, to depart *altogether* from the conventional concept of the nearly-elastic domain and base practical design on "plastic theories"[1], which assume that plane sections remain plane after bending, but which take into consideration the lack of proportionality between stress and strain, and, consequently, employ a non-linear bending stress diagram.

For instance, in Mensch's theory of 1914, which seems to have been the earliest among this group,[2] the stress distribution in the concrete was assumed to be a cubic parabola, while H. Kempton-Dyson, in 1922, used a quarter ellipse.[3] Other investigators, such as Professor V.P. Jensen, advanced the hypothesis that the diagram consists of two straight parts, representing respectively the truly elastic and the plastic domains.[4]

[1] Rhydwyn H. Evans, "The Plastic Theories for the Ultimate Strength of Reinforced Concrete Beams", *J. Inst. C.E.*, London, Dec. 1943, p. 98.

[2] L.J. Mensch, *J. Am. Concrete Inst.*, 1914, Vol. 38, 65.

[3] H. Kempton-Dyson, *Concrete and Constructional Engineering*, 1922, Nos. 5, 6, 7, 8.

[4] Prof. V.P. Jensen, *J. Am. Concrete Inst.*, 1943, Vol. 14, 6, 565.

Quite apart from their intrinsic value, these theories throw also an interesting side-light on the more general problems of the structural safety factor, with particular reference to the numerous modern suggestions tending to depart from the criterion of the "working stress", which for over one hundred years has dominated structural engineering all over the world; and thus, to revert back to the criterion of the "breaking load", as it had been used by the earlier designers in the first half of the last century.

6. Bonded and Non-bonded Types of Pre-stress

Reverting now to the primary subject of the present discussion, i.e. pre-stressed concrete, it must be understood that according to the usual practice, all the structures falling under this title may be divided into two main categories: the bonded and the non-bonded types.[1]

In the first—also described as the "concrete-gripped"—type, the artificial stretching of the wires takes place before concreting starts, by means of special stretching devices. After the casting is completed and the concrete hardens, it adheres to the steel, as in ordinary reinforced concrete, and when the temporary stretching devices are removed, the tensile stress they exerted on the wires is transferred to the concrete, which by preventing the wires from shortening, gets itself compressed, i.e. pre-stressed.

The advantages of the method are that the wires are gripped by the concrete throughout all their lengths, and also, that they are effectively protected against rust. On the other hand, the main disadvantage is that part of the pre-stress is lost when the wires are released from the temporary stretching devices, because, at that moment, the concrete gets compressed, and therefore shortens. The magnitude of this loss may be estimated from the formula obtained by equating the elastic deformations of the two materials (i.e. steel and concrete). Calculations carried out in this manner, show that the loss due to this cause may be as much as thirty per cent of the force originally developed by the stretching devices. In addition to this, comes also the loss of pre-stress due to the shrinkage of the concrete, when it hardens. Experimental figures assessing quantitatively the value of the shrinkage are rather unconclusive, but 0.0005 of the length may be assumed as a rough approximation, and this would correspond to a loss of about 15,000 lb/sq in.

It is, therefore, clear that in structural applications of this method, the original pre-stress, developed by the stretching devices, must be rather high, in order to remain a significant consideration after the completion of the work, and this comes down to saying that carefully manufactured, expensive materials of high strength must be employed, if this method is used.

This method is, therefore, more suitable for mass production in the factory of relatively small elements, for instance, railway sleepers.[2]

On the other hand, when faced with problems relating to long-span concrete bridges, we shall frequently find that the alternative solution, that of the non-bonded, i.e. end-anchored, pre-stressed concrete beam affords usually greater advantages than the bonded, gripped type. The cables, or "tendons", are separated in this case from the surrounding concrete by a sheath or wrapping, but are anchored at the ends, either by fixing them to anchor plates, or by other devices performing the same function. The tendons are, therefore, free to move during stretching, an operation usually carried out after the concrete is cast and has hardened and shrunk.

---

[1] Other types of pre-stress, such for instance as that which consists in using expanding cement concrete, and others, are far less frequent.

[2] This solution is also used for larger works, such as bridges, in the case if the latter are built of a large number of prefabricated small units, by assembling and pre-stressing them in the field.

The loss of the pre-stress such as it is,[1] is therefore much smaller than in the first type, and additional pre-stress is possible, after the dead load is applied and the corresponding deformations have therefore taken place.

These are, evidently, material advantages, but they are partly eclipsed by practical considerations militating against this method, e.g. that the safety of the entire structure depends solely and exclusively on the permanence of the end anchorages; of course, with proper care, this may not be a very serious risk, but from general, common-sense considerations one is naturally inclined to prefer a system in which there are no anchorages to fail. For instance, as in the case with the bonded type of pre-stressed bridges.

Another disadvantage of the end-anchored system is that, since the wires have no solid contact with the concrete (except at their ends) they are exposed to rust and must therefore be protected against corrosion by injecting cement grout into the sheath—a possibly delicate operation adding to the total cost of the work. Though it may be true that such structural problems can be dealt with by the application of a specific technique, they must, nevertheless, be taken account of in comparing the *pros* and *cons* of the two systems.

It is therefore relevant to the present discussion to quote Professor Gustave Magnel—the Belgian specialist on pre-stressed concrete—concerning the requirements which wires used in this type of construction must satisfy.[2] The quotation refers in particular to the conditions prevailing in Belgium, but is applicable by inference to other countries as well.

High tensile strength is a prerequisite. The wires must have an ultimate strength of about 210,000 lb/in² and a conventional elastic limit of 170,000 lb/in². Pre-stressing may then attain:

$$0.6 \times 210,000 = 126,000 \text{ lb/in}^2$$
$$0.8 \times 170,000 = 136,000 \text{ ,, ,,}$$

The wires must be placed in the cable in a predeterminate order and kept in that position during the entire pre-stressing operation. They must be properly aerated; i.e. every wire must be kept at a distance of at least $\frac{3}{16}$ inch from the adjacent wires; for otherwise the grouting of the cable, after pre-stressing, cannot be done properly, and its efficacy in protecting the wires against corrosion and in creating additional bond may be subject to doubt.

The stretching must produce a statically determinate stress distribution among the individual wires, and in order to achieve this objective, not more than two wires must be stretched at a time. If this requirement is not satisfied no guarantee can be given that wires will all carry the same load.

It must be possible to make cables of various sizes, starting with eight wires of 5 millimetres diameter and up to sixty-four wires of 7 millimetres diameter.

Standard stretching apparatus is essential to reduce cost.

Frictional resistances between the wires must be negligible during the stretching operations, and this means using special spacers.

7. Comparison with a Retaining Wall

Since the aim of these pages is to acquaint the student with the basic principles of the

---

[1] At the moment of the transfer of the stress from the stretching devices to the anchorages, the latter yield about 0.05 in to 0.08 in, and this results in a loss of pre-stress.

[2] "Applications of Pre-stressed Concrete in Belgium", Lecture delivered at the Institution of Civil Engineers, London, on Feb. 15, 1949, *J. Inst. C.E.*, April 1949, p. 161. Quoted with small alterations from pages 161 and 162.

PRE-STRESSED BRIDGES

theory of pre-stress, an attempt is made below to outline an analogy between this theory on the one hand, and the common retaining wall on the other.

Let us, then, consider, in the first instance, the retaining wall shown in Fig. 133.

Suppose that the resultant force strikes the section AB at a distance e from the centre. If e is larger than $h/6$, i.e. if this resultant falls outside the middle third (as shown in the diagram), the stress f_t at A will be tension, while the stress f_c at B will be compression. According to the age-old Navier formula, familiar to every technical student, these stresses are expressed by the formula

$$f = \frac{N}{h}\left(1 \pm \frac{6e}{h}\right),$$

or, more explicitly:

$$f_c = \frac{N}{h}\left(1 + \frac{6e}{h}\right) \quad \text{and} \quad f_t = \frac{N}{h}\left(1 - \frac{6e}{h}\right).$$

It will now be shown that the main equations framing the pre-stress theory can be developed from, and are but another presentation of these same, well-known formulae; in fact, let us consider the pre-stressed beam in Fig. 134, which we will suppose to be subject to the effect of the force N developed by pre-stressing, i.e. by stretching the cable, or, to use the modern term introduced by M. Freyssinet, the "tendon".

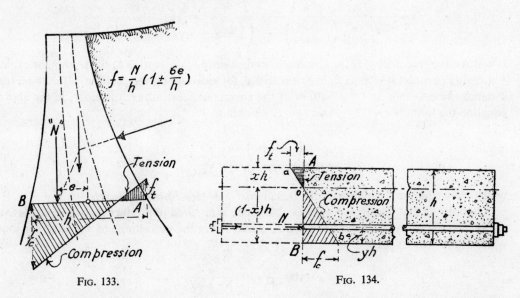

Fig. 133.   Fig. 134.

Our first problem will then be to find (a) the position of the neutral axis, when this force alone is acting, and (b) the magnitude of the force which is necessary to produce a certain compressive stress f_c (required to counterbalance an equal tensile stress due to an external loading, if and when the latter will be applied).

It will be observed that the type of stress distribution in the section AB in Fig. 134, is very much the same as in the section AB, in Fig. 133, viz. we have a normal force N applied at distance e from the centroid of the section, and a stress diagram, which in both cases represents a linear distribution on intensities.

And, therefore, in Fig. 134 as in Fig. 133, the stress intensities will be controlled by Navier's formulae

$$f_c = \frac{N}{h}\left(1 + \frac{6e}{h}\right)$$

$$f_t = \frac{N}{h}\left(1 - \frac{6e}{h}\right).$$

The position of the neutral axis is then determined by the distance xh, measured from the point where the stress is zero to the top of the beam. Since the two triangles, representing in the diagram the distribution of tensile and compressive stresses, are similar, we may write

$$\frac{f_c}{h - xh} = \frac{-f_t}{xh}.$$

Note that the negative sign appears in the right-hand term of this equation because the condition of geometrical similarity, incorporated in the formula, refers, obviously, to the absolute values of the stresses.

Solving:

$$x = \frac{f_t}{f_t - f_c} = \frac{\frac{N}{h}\left(1 - \frac{6e}{h}\right)}{\frac{N}{h}\left(1 - \frac{6e}{h}\right) - \frac{N}{h}\left(1 + \frac{6e}{h}\right)} = \frac{6e - h}{12e} = \frac{1}{2} - \frac{1}{12}\frac{h}{e},$$

in which e, the eccentricity, is the distance from the centre of the section to the tendon, or cable. For design purposes it is more convenient to use, for ascertaining the position of this wire, the distance yh between it and the bottom of the concrete. Substituting, therefore, in the above equation the formula $e = h/2 - yh = h(\frac{1}{2} - y)$ we obtain

$$x = \frac{6\left(\frac{h}{2} - yh\right) - h}{12h(\frac{1}{2} - y)} = \frac{1 - 3y}{3 - 6y} \tag{1}$$

which is a well-known equation in the rectangular pre-stressed-beam theory.

In order to find the stress f_s in the cable which is required to produce a given compressive stress intensity f_c in the concrete, we shall assume that the percentage of the steel per unit width of the beam is p, i.e.

$$p = 100\,\frac{a}{A} = 100\,\frac{a}{h \times 1.00},$$

in which a and A are, as before, the corresponding areas of the steel and of the concrete.

As was the case in the retaining wall shown in Fig. 133, the force N, i.e. the total stress in the wire, $f_s \times a = f_s\,\dfrac{ph}{100}$, is equal to the difference between the total compressive and the total tensile stresses in the concrete, which are represented, respectively, by the area $\frac{1}{2} f_c h (1-x)$ of the triangle ObB, and by the area

$$\tfrac{1}{2} f_t h x = \tfrac{1}{2} f_c h\,\frac{x^2}{1-x}$$

PRE-STRESSED BRIDGES

of the triangle OaA. Hence

$$f_s \frac{ph}{100} = \tfrac{1}{2} f_c h \left[(1-x) - \frac{x^2}{1-x} \right] = \tfrac{1}{2} f_c h \frac{1-2x}{1-x},$$

and therefore

$$f_s = \frac{50 f_c}{p} \times \frac{1-2x}{1-x} \qquad (2)$$

which is the second basic formula of the pre-stressed-beam theory.

Armed with these two formulae, i.e. the equations (1) and (2), we shall now attack the problem of pre-stressed concrete on more general lines, i.e. we shall assume that external forces, such as rolling loads or structural weight, have caused in the upper fibres of the concrete a compressive stress, f_r, and the consequences thereof will be examined.

SECTION II. CALCULATION OF A PRE-STRESSED BEAM

1. The Assumption of Non-cracked Concrete

The case of the unbonded, i.e. end-anchored, beam will be considered as the more *typical* of the two, and the concrete will be assumed non-cracked. For simplicity, the beam will be taken to be rectangular, of unity width. For any other width, b, the formulae are to be multiplied by the constant b. The total forces acting in the section AB (see Fig. 135) will then be as follows:

(a) *Compression in the concrete* represented by the upper triangle in the diagram

$$C_1 = \tfrac{1}{2} f_r h x_1$$

(b) *Tension in the concrete* represented by the lower triangle

$$T_1 = \tfrac{1}{2} f_r \frac{1-x_1}{x_1} h(1-x_1) = \frac{f_r h (1-x)^2}{2 x_1}$$

(c) *Tension in the wires*

$$W_1 = C_1 - T_1 = \frac{f_r h x_1}{2} - \frac{f_r h (1-x)^2}{2 x_1} = f_r h \left(1 - \frac{1}{2 x_1} \right).$$

The argument of these formulae is elementary and does not call for comment.

The position of the neutral axis, viz. the value of x_1, is supposed to be determined, depending on y, p and n. The forces C_1 and T_1 are then represented by the areas of the two triangles of stress appearing in the diagram Fig. 135. Since there are no other normal forces acting in the section considered, except C_1, T_1 and W_1, it is evident that $W_1 = C_1 - T_1$. Thus, all these forces can be visualized as functions of one single parameter, f_r. We can, also, represent the corresponding moment of resistance as a function of this same parameter; in fact, this moment can be calculated as the sum of the moments of T_1 and W_1 about C_1, and we shall thus obtain:

$$M_r = \frac{f_r h^2}{6 x_1} (3y - 1 + 3x_1 - 6x_1 y). \qquad (3)$$

It will be observed that thus far, we have left the choice of the value of y entirely free. It must be realized, however, that for reasons of strength, efficacy and economy, this value should be as small as structurally feasible, i.e. the cable, or tendon, must be located as near as possible to the border on the extended side of the beam. For instance, we may choose y equal to 1/10. Substituting, then, this value in the formulae (1), (2) and (3), we shall obtain:

$$x = \frac{1 - 3 \times 0.1}{3 - 6 \times 0.1} = 0.29 \tag{1a}$$

$$f_s = \frac{50 f_c}{p} \times \frac{1 - 2 \times 0.29}{1 - 0.29} = 29.5 \frac{f_c}{p} = 72.1 \frac{f_t}{p} \tag{2a}$$

$$M_r = f_t h^2 \left(0.4 - \frac{0.11}{x_1} \right). \tag{3a}$$

Note that in the foregoing investigation we have considered the local effect only, of the stress f_r, which varied throughout the length of the beam depending on the bending moment in every section thereof.

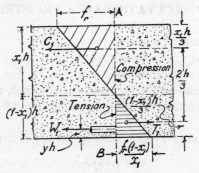

FIG. 135.

Apart from that there is, however, an additional effect also, consequent on the overall extension of the wire; or, in other words, since owing to the deflection of the beam the distance between the extreme points in which the wire is anchored increases, the wire is, consequently, extended. It produces, therefore, an elastic reaction on the concrete, similar in a sense to that of the initial pre-stress force N, which however, was due to an artificial stretching of the cable.

Hence, we must take into account two superimposed states of stresses: (a) the variable effect we have already estimated, and (b) that of an additional stress, f'_s, in the wire, which is constant for all the length of the beam. The (b) problem remains therefore, to be solved.

Taking as basic parameter this additional stress in the wires, f'_s, we calculate from formulae (2) and (2a) the corresponding additional compressive stress in the concrete:

$$f'_s \frac{p}{50} \left(\frac{1-x}{1-2x} \right) = f'_s \frac{p}{50} \left(\frac{1-0.29}{1-2 \times 0.29} \right) = \frac{p}{29.6} f'_s. \tag{4}$$

Since the triangles representing in the diagram the tensile and compressive stresses are similar, we can calculate the maximum additional tensile stress in the concrete, multiplying this equation by the ratio $x/(1-x) = 0.29/(1-0.29) = 1/2.45$. We thus obtain

$$\frac{p}{72.5} f'_s.$$

On the other hand, if formula (4) is multiplied by the ratio $(1-x-y)/(1-x) = 1/1.164$, we

PRE-STRESSED BRIDGES

shall find in the same manner, the corresponding compressive stress for the concrete adjacent to the wires, viz.

$$f_a = \frac{p}{34.4} f'_s. \tag{4a}$$

Combining these various secondary effects with the original stresses shown earlier in Fig. 135, the final diagram representing the stresses in the concrete is of necessity a symmetrical pattern, because after excluding the stress in the cable, the internal stresses must be balanced by the external moment, for there are no external forces parallel to the beam.

Hence, the total stress at the top and bottom of the beam is

$$f_T = f_r + f'_s \times \frac{p}{72.5}.$$

For the concrete adjacent to the wires we multiply this formula by the reduction coefficient $(0.5-y)/0.5 = (0.5-0.1)/0.5 = 0.8$, and thus find:

$$0.8 \left(f_r + f'_s \times \frac{p}{72.5} \right).$$

So long, however, as the basic parameter, i.e. the additional stress in the tendon, f'_s, appearing in these formulae, remains undetermined, the solution is not complete.

The calculation of the stress f'_s is a statically indeterminate problem, i.e. its value will be found by equating the total strain of the wires to the total strain of the concrete (between the anchorages).

Note, however, that since the beam we are investigating is of the non-bonded, i.e. post-tensioned, type, the wires are supposed to have no rigid contact with the concrete except at their ends, and the stress in them is, therefore, constant over all their lengths. Hence, the corresponding strain can be easily determined, being as it is equal to $f'_s L/E_s$, in which L is the length of the beam.

The problem for the strain of the concrete is however more involved, for in this case the stress changes over the length of the span, depending, as it does, on the form of the bending-moment diagram. The solution must, therefore, include a "form-factor", γ, representing the ratio of the average to the maximum bending moments in the span. The diagrams in Fig. 136 will help to visualize the physical significance of this factor.

For a uniformly distributed loading $\gamma = 2/3$; for a single concentrated load $\gamma = 1/2$, and so on. Using, then, this coefficient, the average stress in the concrete adjacent to the wires is

$$0.8 \left(f_r + f'_s \times \frac{p}{72.5} \right) \gamma - \frac{p}{34.4} f'_s.$$

Note that the factor γ is included in the first term of this formula only, because the second term embodies the effect of the normal force developed by the tendon, which remains the same for all the length of the beam.

The corresponding strain is obtained by multiplying this expression by the length of the beam, L, and dividing the product by E_c. Equating the result thus obtained to the strain of the steel, which was found earlier, and multiplying the left-hand and right-hand terms of the formula by E_s/L, the hyperstatic equation is found to be as follows:

$$f'_s = n \left[0.8 \left(f_r + \frac{p}{72.5} f'_s \right) \gamma - \frac{p}{34.4} f'_s \right].$$

Solving this with reference to f'_s

$$f'_s = \frac{72.5\gamma}{\dfrac{90.6}{n} - p(\gamma - 2.64)} f_r \qquad (5)$$

in which, as before, $n = E_s/E_c$. Adding this to the stress due to pre-stress, as per formula (2a), the total stress in the wire is:

$$\frac{72.5\gamma}{\dfrac{90.6}{n} - p(\gamma - 2.64)} f_r + \frac{72.1}{p} f_t.$$

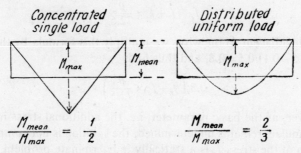

Fig. 136.

On the other hand, to obtain the value of x_1 for the position of the neutral axis in Fig. 135, we equate the moment of resistance given by (3) to the equivalent moment of the comprehensive symmetrical diagram referred to on p. 151 which is obviously equal to $\dfrac{h^3}{6} f_r$, and thus obtain

$$x_1 = \frac{3 f_r y - f_r}{f_T - 3 f_r (1 - 2y)} = \frac{0.7 f_r}{1.4 f_r - f'_s \dfrac{p}{72.5}} = \frac{1}{2 - \dfrac{1.43 p\gamma}{90.6/n - p(\gamma - 2.64)}} . \qquad (6)$$

It will be remembered that we have assumed in the foregoing that the concrete was not cracked, and we have, therefore, introduced in the relevant formulae the entire section of the beam, without any deduction for cracking, which would have been *de rigueur*, had we been dealing with an ordinary reinforced concrete beam.

2. Effect of Cracks

This, indeed, is the correct way to design pre-stressed concrete, for as soon as cracking begins, we are getting the structural type, which Freyssinet describes as "discontinuous" (see p. 143), and the main advantage of pre-stress is, then largely lost. But, such considerations must not deter us from investigating the state of stress which would prevail *if* cracks did actually occur. In fact, we shall show in the following, that should such conditions indeed occur, the cable would take more than its allotted share of the stress, and this would naturally result in a rapid rise of the rate of deformations, as predicted by M. Freyssinet.

Let *mn* in Fig. 137 (*a*) be the net diagram of pre-stress after deduction of all losses, and let *ap* in Fig. 137 (*b*) represent the stresses due to external loads—also, after deduction of losses.

PRE-STRESSED BRIDGES

Subtracting, then, the abscissae of the former from those of the latter, we obtain the actual diagram of the net stresses in the section OO, as shown by cd. Note that the tensile breaking stress of the concrete is assumed to be f_0.

The abscissae of diagram cd are then equal to $f_r - f_t$ at the top, and $f_r(1-x_1)/x - f_c$ at the bottom. If this line intersects the vertical ii, which is drawn at a distance f_0 from the axis and which represents, therefore, the limit of the tensile resistance of the material, the concrete will crack, and all the part of the beam located below the point of intersection e will cease to resist the bending moment, with the result that a larger load must then be taken by the cable. The point of this argument is obvious enough, but our next problem is to estimate quantitatively the magnitude of this effect, in order to get a clear idea of the true importance of the cracking on the bearing capacity of the beam. Note that in the diagram, this additional force taken by the wires is represented by the area of the shaded trapezium $efoc$.

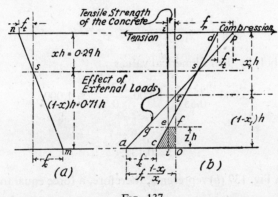

Fig. 137.

To solve this problem we must assume certain values of the variables appearing in the formulae, similar to those met with in common practice; for instance as follows:

(a) Compressive stress due to bending $\qquad f_r = 13{,}000$ lb/in²
(b) Compressive stress due to pre-stress $\qquad f_c = 8{,}000$,, ,,
(c) Tensile strength of concrete $\qquad f_0 = 1{,}000$,, ,,
(d) Percentage of wires of reinforcement $\qquad p = 2$
(e) Form-factor for the bending moment diagram $\qquad \gamma = 0.67$
(f) Distance from centre of reinforcement to bottom of beam $\qquad yh = 0.1h$

Our first objective is the force represented by the shaded trapezium in the diagram Fig. 137 (b). The top and bottom of this trapezium are known—they are f_0 and $f_r \dfrac{1-x_1}{x_1}$ respectively. But its height, zh, remains still to be determined.

With this aim in view, we begin by finding x_1 from formula (6), viz.

$$x_1 = \frac{1}{2 - \dfrac{1.43 \times 2 \times 0.67}{90.6/15 - 5(0.67 - 2.64)}} = 0.55.$$

This yields for the bottom of the trapezium the value

$$f_r \frac{1-x_1}{x_1} - f_c = 13{,}000 \frac{1-0.55}{0.55} - 8{,}000 = 10{,}630 - 8{,}000 = 2{,}630 \text{ lb/in}^2.$$

Note that had the result of this last calculation been less than $f_0 = 1,000$ lb/in², no cracking would have occurred at all.

From the triangle *asc* in Fig. 137, we get:

$$ge = f_c \frac{(1-x-z)h}{(1-x)h} = 8,000 \frac{0.71-z}{0.71} \text{ lb/in}^2.$$

On the other hand, from the triangle *aot*

$$gf = f_r \times \frac{1-x_1}{x_1} \times \frac{(1-x_1-z)h}{(1-x_1)h} = 13,000 \times \frac{0.55-z}{0.55} \text{ lb/in}^2.$$

The difference between *gf* and *ge* is the tensile strength of the concrete, f_0, represented by *ef*. Thus, the equation for finding z is

$$f_r \frac{1-x_1-z}{x_1} - f_c \frac{1-x-z}{1-x} = f_0.$$

Replacing the symbols by their numerical values

$$13,000 \frac{0.45-z}{0.55} - 8,000 \frac{0.71-z}{0.71} = 1,000.$$

Solving

$$z = 0.132.$$

The shaded area in Fig. 137 (*b*) represents, therefore, a force equal to

$$\frac{z}{2}\left(f_r \frac{1-x_1}{x_1} - f_c + f_0\right) = \frac{0.132}{2}(2,630 + 1,000) = 240h \text{ lb},$$

and, since the sectional area of the cable is $ph/100 = 0.02h$, the corresponding increase of the unit stress is

$$\frac{240}{0.02} = 12,000 \text{ lb/in}^2,$$

which is by no means negligible.

It should be realized that this is a sufficiently accurate, *but not a precise*, result, for in order to obtain the true solution, the neutral axis in Fig. 137 must be re-calculated and the stress in the wire re-determined, bearing in mind that the cracks will affect part of the beam only, and, in turn, will therefore reflect of the value of γ.

The method used in this calculation is, however, fully adequate to show that beyond a certain, critical limit the increase in the load causes a discontinuity in the stress-strain relation (in the sense of the term used by M. Freyssinet, see p. 143), which means that in pre-stressed structures, the safety factor should be expressed preferably in terms of loads rather than in terms of stresses. It should be realized, however, that this approach to the problem of the safety factor must not prevent us from verifying that under normal working conditions, the stresses are kept within reasonable limits.

Better to set out the point about the effect of the cracking of concrete in prestressed beams, Fig. 138 represents the values of z plotted as ordinates, against the values f_r as abscissae.

PRE-STRESSED BRIDGES

This is a graphic demonstration of the fact that as load increases, wires, apart from a proportionally increasing stress, normally allotted to them, will also carry a gradually augmenting part of the stress, which was previously taken by the concrete. It also shows that the development of these cracks depends (but not to a great extent) on the percentage of the steel. The depth of the section and the pre-stress are more effective in preventing crack occurrence.

It must be realized, however, that, quantitatively, the diagram is subject to the same reservations as the method employed in preparing it.

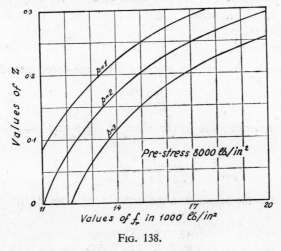

FIG. 138.

3. Effect of "Creep"

It remains now to consider the effect of the creep, which has caused so much trouble in the earlier years of the history of pre-stressing. In fact, had creep been a permanent condition, pre-stress would have been a temporary, i.e. transient state, and as such, must have been written off as a practical engineering proposition. It is, therefore, fortunate that both experiments and practice show that the slow contraction of concrete under a constant stress—which we call "creep"—decreases in course of time and attains probably its greatest value after twelve months. After which period, the volume occupied by the material remains, for all practical intents and purposes, constant.

A design engineer may form a fairly clear idea about this phenomenon, if he compares it with the settlement of an earthen bank, which takes place chiefly during the few months after the work is completed, but thereafter dies gradually out.

What is essential is, therefore, to ascertain how much of the original pre-stress will remain permanently effective, after the contraction of the concrete (due to creep) has taken place.

In investigating this problem we must take into account the results of experimental evidence, which show that creep is accompanied by a drop in the value of the Young's Modulus, and that the importance of this effect depends on the crushing strength of the material. For instance, for a concrete with a crushing strength of 5,000 lb/in² the modulus may be reduced after twelve months to 1/2.5 of its original value, whereas for a weaker material with a crushing strength of only 2,000 lb/in², the reduction may be as much as 1/3.

Here again we shall therefore use the hyperstatic equation, incorporating the principle that the deformations of concrete and steel must be the same; i.e., that as the concrete contracts, the length of the wires shortens in the same proportion. In terms of elasticity this means that the additional strain, due to creep, must be equal in both cases.

The following notation will be used:

Original stress in concrete, adjacent to wires	f_a
,, ,, *in steel, before creep occurs*	f_s
Reduction in the stress f_a due to creep	Δf_a
,, ,, ,, ,, f_s ,, ,, ,,	Δf_s
Young's Modulus for concrete before creep took place	E_c
,, ,, ,, ,, *after* ,, ,, ,,	E_c/k
Modular ratio before creep	n
,, ,, *after creep*	kn

With this notation, the strain in the concrete in the pre-creep period is

$$\frac{f_a}{E_c} = \frac{f_a}{E_s/n}.$$

The change of this strain, due to the effect of creep and reduction in the modulus of elasticity, is

$$\frac{f_a - \Delta f_a}{E_s/kn} - \frac{f_a}{E_s/n}.$$

This must be equal to the decrease in the strain in the wires $\Delta f_s/E_s$. Hence the hyperstatic formula

$$\frac{\Delta f_s}{E_s} = \frac{f_a - \Delta f_a}{E_s/kn} - \frac{f_a}{E_s/n}$$

which reduces to

$$\Delta f_s = (k-1)nf_a - kn\Delta f_a.$$

This equation contains two unknowns, Δf_s and Δf_a. In order to solve the problem we need another equation. This may be, for instance, formula (4a) on p. 151 *ante*

$$f_a = \frac{p}{34.4} f_s.$$

Solving, then, these simultaneous equations:

$$\Delta f_s = \frac{n(k-1)}{1 + \frac{knp}{34.4}} \times f_a \quad \text{and} \quad \Delta f_a = \frac{np(k-1)}{34.4 + knp} \times f_a.$$

Suppose, now, that $f_a = 5{,}000$ lb/in² and that the corresponding values of n, k and p are 10, 2.5 and 1.5. We shall then find:

$$\Delta f_s = \frac{10(2.5-1)}{1 + \frac{2.5 \times 10 \times 1.5}{34.4}} \times 5{,}000 = 35{,}900 \frac{\text{lb}}{\text{in}^2}.$$

Hence

$$\Delta f_a = \frac{1.5}{34.4} \, 35{,}900 = 1{,}560 \, \frac{\text{lb}}{\text{in}^2} \quad \text{and} \quad \frac{f_a - \Delta f_a}{f_a} = 69 \text{ per cent.}$$

Thus, about two-thirds of the original pre-stress will remain effective after the creep has taken place. On the other hand, for a weaker material, i.e. for $f_a = 2{,}000$ lb/in², $n = 20$ and

PRE-STRESSED BRIDGES

$k=3$, the reduction of the pre-stress in the concrete will be found (using the same formulae) equal to 964 lb/in^2, which means that only about half of the pre-stress will indeed be permanent.

Professor A.L.L. Baker, who is primarily responsible for the method and examples of the foregoing pages, remarks[1] that the result of this calculation exceeds the figures published in the reports on experiments, the empirical values for the loss ranging from 5,000 to 10,000 lb/in^2. He suggests that the difference may possibly be due to the high rate of pre-stress incorporated in the calculations.

SECTION III. SKEW SPANS

1. General Considerations

It must be realized, however, that the case thus far considered—that of the rectangular beam—although admirably suited for a general explanation and understanding of the pre-stress theory, is not the one which is most likely to be met with in practice. In fact, should pre-stressed beams be used for the current type of road bridges, or canal and railway crossings, this will most probably occur with skew bridges, for this is the particular type in which wider spans may be of practical advantage, even when the object to be bridged over is relatively narrow.

Consider, for instance, the case shown in Fig. 139. In order to reduce the cost of the reinforced concrete deck, the main girders might have been placed at right angles to the axis of the bridge. This can be achieved either by diverting the alignment of the roadway, as in (a), or by building a wider deck, as in (b).

Both solutions have been attempted, but they are not always suitable in practice; the diversion (a), because the cube of the earth-work required to carry out such a project is relatively too large, and the expropriations are expensive; also, the S-shaped road alignment is objectionable from the standpoint of traffic facilities. On the other hand, with the (b) alternative the additional lengths, m and n, of the abutments are a costly item of the estimate.

There remains, therefore, the (c) solution only. For a straight bridge of the same span, an intermediate pier, as shown by the dotted lines, would have yielded the best solution, but such a pier blocks the clear opening to a too great extent, and is, therefore, subject to adverse criticism.

We are, consequently, left with the only alternative of a skew-bridge of a span somewhat wider than that of the usual type.[2]

Note that here a minor work may become a major design problem by reason of the frequency of its recurrence.

2. Examples

From among the many structures produced as examples in the rather voluminous modern

---

[1] A.L.L. Baker, *Reinforced Concrete* (London, Concrete Publications Ltd., 1949), p. 178.

[2] For the particular characteristics of this skew span as such, the reader's attention is directed to a valuable paper on the subject by Maurice Barron, "Reinforced Concrete Rigid Frame and Arch Bridges", in *Trans. Am. Soc. C.E.*, Vol. 116, 1951, pp. 999 *et seq.*, which contains useful information for the designer. Suggestive ideas will also be found in the following discussion.

158   ARCHES AND BRIDGES

literature on pre-stressed concrete, two skew-bridges over British railways[1] are selected for quotation below, as typical for their class. The girders employed in remodelling the superstructure of these two bridges were of the pre-cast pre-stressed concrete-gripped (bonded) type. They were taken from the stock stored by the British Ministry of War Transport for emergency bridge construction during the war, and were transported to the spot complete and erected cheek-by-jowl, thus covering the entire width of the roadway (28 girders in one bridge and 32 in the other).

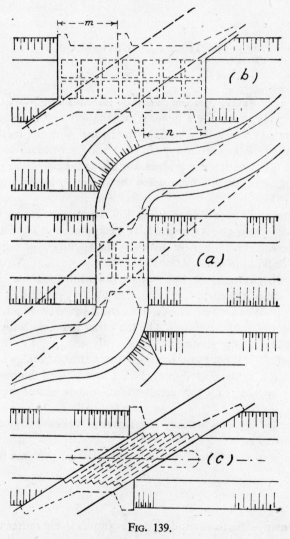

Fig. 139.

In the bridge over the London, Midland and Scottish Railway the girders were 44 feet long, of box-shaped cross-section $18\frac{1}{2}$ in by 27 in deep (see Fig. 140 (a)) and weighed 7 tons each. The pre-stressed reinforcement consisted of 28 hard steel wires (90–100 tons) of 0.2 in

[1] Alan Andrew Paul, "The Use of Pre-cast Pre-stressed Concrete Beams in Bridge Deck Construction", *J. Inst. C.E.*, London, Nov. 1943, p. 19.

PRE-STRESSED BRIDGES

diameter each, but in addition to this, there was also a non-pre-stressed mild steel reinforcement, required to resist excessive tension developed in the upper part of girder due to pre-stress. Twenty-four inch solid cores, at each end of the girder, served as bearing panels and anchorage blocks. A reinforced concrete deck slab was placed above the girders (about 4 inches deep).

In the bridge over the London and North Eastern Railway (skew-span 45 feet, 10 inches) the beams were of an H-section (see Fig. 140 (b)), 12 in wide by $27\frac{1}{2}$ in deep. The pre-stressed reinforcement consisted of thirty wires of the same type as in the first bridge. The weight of these beams was $5\frac{1}{2}$ tons each. The deck slab was 4 inches minimum and was turned down along its edges, as it was observed on the first bridge that the vibrations from the rail traffic caused the outside beams to break away.

In working out the programme for carrying out such works it is essential to ensure that at no time will the girders be inversed. In fact, so long as these girders are in their correct position, the pre-stress neutralizes, completely or partly, the stresses due to structural weight, but should the girder be inversed, accumulation of stress will take place, and the material may fail. It is, therefore, good practice to provide the girders, when casting them, with special hooks, which are used when lifting, transporting and depositing them in position.

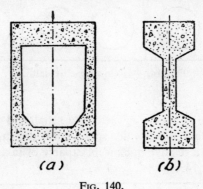

Fig. 140.

Comparing the box type shown in Fig. 140 (a) with the H-type in Fig. 140 (b), Mr. Paul, the author of the quoted article, concludes in favour of the latter class,[1] which is in full accord with the modern trend of structural tendencies in general.

SECTION IV. POPULAR TYPES FOR PRE-STRESSED BRIDGES OF MODERATE SPAN

1. Solid Slab

In general, in deciding which type of short pre-stressed bridge is most suitable for use in any particular case, the designer will be guided by the following overall considerations.

The simplest and most obvious type for a pre-cast bridge for the smallest span is, of course, the solid pre-tensioned slab, as shown in Fig. 141, with various kinds of surfacing above it.

The average section of such slabs is about 1.0×0.3 metres. The bending moment, for which they are to be designed, is a function of the square of span, whereas the shear depends

---

[1] *Loc. cit.*, p. 28.

chiefly on the span and live load. The structural weight, in this case, is relatively non-important, and therefore, it is quite reasonable to use here a solid beam.

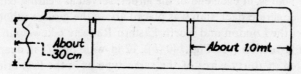

Fig. 141.—Solid pre-cast slab type.

2. Hollow Slab

However, as the span of the bridge increases, the dead weight of the beam becomes a consideration, and steps must then be taken to get a lighter structure. We are thus led to the concept of a "hollow" pre-tensioned slab, for instance as shown in section in Fig. 142. The average section, for which this type is suitable, is about 1.0×0.45 metres.

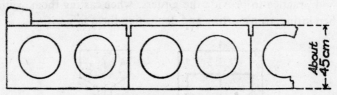

Fig. 142.—Hollow pre-cast slabs.

3. I-beams

If we follow the same line of reasoning further, we shall find that the next type, quoted as a convenient solution for bridges of still wider spans, is the pre-stressed I-beam, with wide top and bottom flanges, and with a relative thick web; capable of taking a still considerable shear. This type is illustrated in Fig. 143.

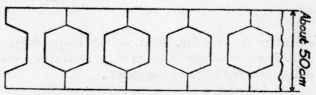

Fig. 143.—Pre-cast I-type girders.

4. Channel-slab and Hollow Stringer

Next, we should mention a rather wide variety of types, which are frequently used by designers, for somewhat wider spans; but, which still belong to the category of "narrow"

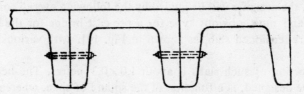

Fig. 144.—Pre-cast channel-slab bridge.

PRE-STRESSED BRIDGES

bridges. The selection of the particular type to be selected for use in such cases, depends largely on the personal "like" or "dislike" of the author of the project. From among these types we may quote the "channel-slab bridge" shown in Fig 144, the "hollow stringer" type appearing in Fig. 145, and so on.

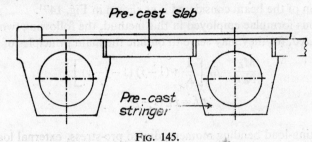

FIG. 145.

5. Combination of Pre-cast Elements with Cast-in-situ Parts

The above examples illustrate what may possibly be regarded as the maximum limit for a narrow span bridge, but in order to widen the scope of our discussion, we may also quote a "moderate" type. The particular, design of such bridges varies in rather wide limits, but their

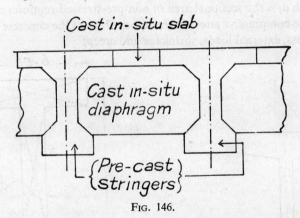

FIG. 146.

general arrangement remains more or less the same; and namely, as shown in Fig. 146, it consists of strong pre-cast stringers, with a wide upper flange, and of cast-*in-situ* diaphragm and narrow slab.

Apart from the above, there are several special types of pre-stress girders, particularly recommended by certain authors. One of such types is described in the following section.

SECTION V. GIFFORD TYPE

In designing pre-stressed beams, advantage may be taken of the method and tables presented in two papers on the subject by Mr. Frederick William Gifford.[1] In the first of these papers the design is based on *working* loads, but in the second Mr. Gifford adopts the ideas

---

[1] "The Design of Simply Supported Prestressed Concrete Beams for Working Loads", *Proc. Inst. C.E.*, London, December 1953, p. 589, and "The Design of Simply Supported Prestressed Concrete Beams for Ultimate Loads", *ibid*, April 1954, p. 125.

briefly outlined on pp. 143 and 144, *ante*, and gives a solution based on the ultimate loads *at failure*. To master the solution in all its various aspects the reader is referred to the original publications. The following brief information gives a general idea of the method of the first paper only.

The type section of the beam considered is as shown in Fig. 147[1].

From the various formulae employed in this method, the following two equations call for being particularly cited, for they may serve to outline the main principles of the solution:

$$\frac{M_L}{b_a h^2} = f_{iL}\left[\frac{1}{K} + (1-j)(1-j-y)\right] \qquad (1)$$

and
$$p f_s = f_{iL}(1-j) \qquad (2)$$

in which

M_L is the full working-load bending moment, due to pre-stress, external loads, creep, shrinkage, etc.;

h is the overall depth of the section (as before);

b_a is the average width of the section, calculated by dividing the net area of the concrete section, A_1, by the overall depth, h, i.e. $b_a = A_1/h$;

A_1 is the area of the net concrete section, i.e. the area, A, of the section of concrete proper, plus na_1, in which a_1 is the sectional area of non-pre-stressed reinforcement;

f_{iL} is the permissible compressive stress in the upper fibres of the concrete under full working load, i.e. pre-stress, external loads, shrinkage and creep;

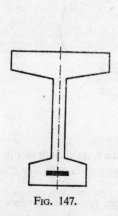

Fig. 147.

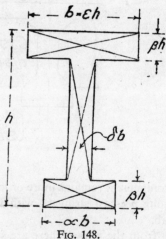

Fig. 148.

K is the shape factor, i.e. $\dfrac{b_a h^3}{I_1} = \dfrac{A_r h^2}{I_1}$ in which I_1 is the second moment of the net concrete area, A_1, about its centroid (this factor varies in practice from 6 to 12);

p is the proportion of pre-stressed steel a_s/A_1;

jh is the depth of the centroid of the area, A_1, below the top of the section;

yh is (as before) the distance from the centroid of the pre-stressed wires to the bottom of the section;

f_s is the stress in these wires.

[1] Note that this is somewhat similar to the main girder in Fig. 146.

PRE-STRESSED BRIDGES

Abstract from F. W. Gifford's Table

	δ	0	0.1	0.2	0.3	0.4
$\alpha=1.0$	j	0.500	0.500	0.500	0.500	0.500
$\beta=0.25$	K	6.87	7.45	8.01	8.56	9.09
$\alpha=0.5$	j	0.375	0.390	0.401	0.411	0.419
$\beta=0.25$	K	7.69	8.41	9.12	9.81	10.46
$\alpha=0.2$	j	0.250	0.286	0.313	—	—
$\beta=0.25$	K	12.0	12.20	12.60	—	—
$\alpha=1.0$	j	0.500	0.500	0.500	0.500	0.500
$\beta=0.15$	K	5.49	6.43	7.28	8.06	8.77
$\alpha=0.5$	j	0.358	0.392	0.413	0.427	0.437
$\beta=0.15$	K	6.12	7.29	8.32	9.21	10.00
$\alpha=0.2$	j	0.217	0.296	0.341	—	—
$\beta=0.15$	K	9.82	9.88	10.52	—	—
$\alpha=1.0$	j	0.500	0.500	0.500	0.500	0.500
$\beta=0.1$	K	4.92	6.23	7.33	8.24	9.01
$\alpha=0.5$	j	0.350	0.402	0.428	0.442	0.452
$\beta=0.1$	K	5.54	7.06	8.32	9.29	10.11
$\alpha=0.2$	j	0.200	0.320	0.371	—	—
$\beta=0.1$	K	8.93	9.02	9.90	—	—

Note that under the conditions qualifying M_L and f_{tL}, the stress, f_{bL}, in the bottom fibres of the concrete, is assumed to be zero.

The equations (1) and (2) can be used for finding the section of the beam, provided K and j are known. To expedite this calculation, we consider a schematized flanged section, consisting of three rectangles, as shown in Fig. 148.

The depth, h, and the four dimensionless coefficients, ε, α, β and δ, suffice to determine all the dimensions of such a section. They can, therefore, be used to find the values K and j appearing in the equations.

Mr. Gifford has calculated a table of these values, an abstract of which is reproduced on p. 163. This abstract is intended to explain the method only, of the table, but for design purposes the original should be consulted.

In designing a section we first assume $\alpha = 1.0$ and choose reasonable values of β and δ; for instance, for small beams $\beta = 0.25$ and $\delta = 0.25$ to 0.40. For larger works we may start with $\beta = 0.1$ and $\delta = 0.1$ to 0.25. The design proceeds then by successive approximations.

One should not forget that the foregoing explanation is only an outline of Mr. Gifford's method. In actual design, in addition to M_L, we consider also the additional sustained-load moment and the instantaneous stresses produced thereby, at the time of transfer of the stress to the concrete, and other structural characteristics of the case, which will be clear from studying the original publication.

Note that the depth of the section is an important consideration, particularly in the case when the height of the road-level on the bridge is too small to permit using a girder of normal, i.e. economical, depth. We must, then, either raise the roadway level artificially, i.e. remodel the road approaches to the bridge, or use specially designed, particularly low girders, which, though not the most economical solution in themselves, will nevertheless allow to save the expenditure on costly road remodellings. Viewed from this standpoint, the economical aspect of the design problem bearing on the height of the pre-stressed I-girders, is a possibly significant point.[1] There is, however, a limit beyond which the height of the girders cannot be reduced. It is fixed from considerations bearing on the maximum specified ratio of deflection to span (usually 1/800).

Assume, for example, a uniformly distributed live load P (total for span). The maximum deflection is then

$$\delta = \frac{P}{EI} \frac{5l^3}{384}$$

and the stress is $f = Mi/I = Pli/8I$, in which i is the height of the centroid of the section above its bottom. From these two equations

$$\delta = \frac{5}{48} \times f \times \frac{l^2}{Ei}.$$

It follows that the ratio of span to height of beam is

$$\frac{l}{h} = \frac{48}{5} \times \frac{\delta}{l} \times \frac{E}{f} \times \frac{i}{h}.$$

Substituting in this $\delta/l = 1/800$ and assuming i/h about 2/3, Professor Robertson[2] obtains, for the minimum permissible height, h, the formula

$$\text{max.} \frac{l}{h} = \frac{8}{1{,}000} \times \frac{E}{f}.$$

Hence, whatever the economical considerations, the girder cannot be made shallower than given by this formula.

Another way of looking at the same problem is based on the oscillation period. This is approximately equal to (deflection in feet)$^{1/2}$, the deflection being calculated ignoring the pre-stress.

---

[1] In this connection attention may be called to a paper by Professor Reginald George Robertson on "Prestressed Concrete Beams: the Economical Shape of Section", published in *Proc. Inst. C.E.*, London, April 1954, p. 242.

[2] *Loc. cit.*, p. 224.

In concluding this outline of the potential scope of application of the theory and practice of pre-stress to small-span bridges, it seems appropriate to quote a few abstracts from M. Freyssinet's catechism on the subject; for instance:[1]

"A compression exerted by a tendon, fixed to the element compressed, so that *both deflect together*,[2] can never produce *buckling*,[2] even if carried as far as rupture by compression."

"There exists a substantial difference in the nature and consequences, insofar as the factor of safety is concerned, between a stress in concrete produced by a stretched tendon, and a stress of the same numerical value caused by a load. One can *and should*[2] permit, in the former case, values[3] 50 and even 100 per cent higher than in the latter."

"The transmission of the forces exerted by the anchorage to the concrete must follow a perfectly clear and clearly defined path, every link of which must be capable of being checked. An efficient distribution of anchorage forces over the concrete is the most reliable criterion of the value of a scheme and has much more importance than a vain and illusory exactitude in the intensities of pre-stress."

---

[1] *Ut supra*, p. 345.
[2] Italics by writer.
[3] This means probably "stress limits".

Chapter Six
SPECIFICATIONS FOR STEEL BRIDGES

SECTION I. EARLIER HISTORY

1. Foreword

The steel bridges we have so far been considering, i.e. the rolled-joist type (see Chapter Three) were of so simple a design that the question of permissible stress was of relatively little importance, as compared with the various empirical rules and practical considerations which constituted the main and significant element of the design.

As we are now concerned with a somewhat more elaborate class of work; the permissible-stress problem must be examined more carefully, bearing in mind that this examination will refer not only to bridges alone, but also to other classes of steel-work, such as gates of various descriptions, penstocks, gantries, steel syphons, aqueducts, etc.

Even in the case in which the metal equipment for a structural work is designed and supplied by a contracting firm, it is nevertheless *the civil engineer's duty* to prepare the relevant specifications and other contract documents; and what is often done in such cases is to copy from an earlier contract the technical stipulations for the metal-work—a most objectionable practice which cannot fail to lead to "mental paralysis".

Without necessarily possessing highly specialized knowledge on the subject, a civil engineer must, nevertheless, be capable of understanding the reasons for adopting different stresses in various parts of a metal structure, whatever the latter may be. It is therefore quite obvious that unless the bases of the relevant theory are known, there is no room for a deeper insight and reasonable advance, or improvement.

2. Intensities of Working Stresses in General

The rational selection of permissible intensities of working stresses is possibly one of the most fascinating problems in the practical life of a design engineer, since the most elegant calculations and advanced methods of stress-analysis are of no practical avail, unless the relevant permissible working stress is judiciously chosen.

The various working limits which have to be considered in the usual steel-design are: the tension, the compression, the shear, the bearing, and sometimes, the torsion. In the current form of specification, the limits for the latter four types of stresses are given as a proportion of the limit for the first type, i.e. tension.

Hence, it is convenient to divide the discussion into two parts, namely:

(*A*) Permissible tensile stress.

SPECIFICATIONS FOR STEEL BRIDGES

(B) Permissible limits for compression, shear and bearing—all of them as functions of the limit for tension.

3. Permissible Intensity for Tensile Stress

In order to obtain a sufficiently clear picture of this subject, it must of necessity be presented in its historic perspective.

The first wrought-iron riveted bridges of the beam type were built in England, in about the middle of the last century. For instance, the "Britannia" tubular bridge (see Figs. 149 and 150) was begun in 1846. But the principle of proportioning the members of a bridge according

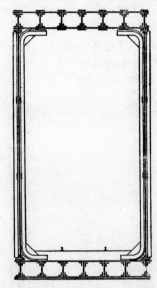

Fig. 149.—Cross-section of the Britannia Bridge, 1849.

to the calculated stresses was introduced, in England, only towards 1860. The safety and the strength of the earlier designs were estimated with reference to the ultimate breaking load, as opposed to the working stress. In fact, in the two or three decades ending in 1859, the contemporary Board of Trade regulations stipulated that the breaking load of all wrought-iron bridges should be six times the prescribed *maximum rolling load*.

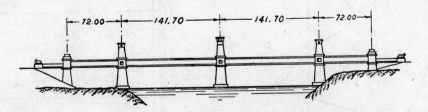

Fig. 150.—Elevation of the Britannia Bridge.

The calculation of the breaking load was very rudimentary; in fact, the formula used was:

$$W = C\frac{at}{l}$$

where W was the breaking load in tons,
a was the area of the flange in tension,
t was the overall depth,
l was the span of girder between the supports.

The coefficient C was estimated empirically, from experiments on models representing the bridge to a fairly large scale. Thus for instance, the model of the "Britannia" bridge was one-sixth of the full size of the prototype (largest span 140 metres). On the average, the coefficient C was 80 for tubular bridges, 74 for plate girders, and 67 for lattice girders.

Fairbairn, who assisted Stephenson in the design of the "Britannia" bridge, writes:

"Owing to the success of these undertakings (the "Britannia" and Conway bridges) there was a general demand for wrought iron bridges in every direction, and numbers were made.... The defects and breakdowns which followed the first successful application of wrought iron to bridge building led to doubts and fears on the part of engineers.... Ultimately it was decided by the authorities of the Board of Trade, but from what data I am

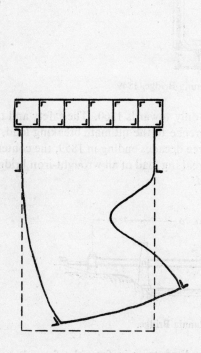

Fig. 151.—Failure of model bridge tested in 1847 by Hodgkinson.

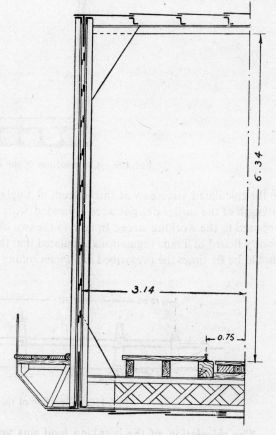

Fig. 152.—Cross-section of Nogat Bridge at Marienburg.

not informed, that no wrought iron bridge should, with the heaviest load, exceed a strain of 5 tons per square inch."

The resistance of iron against static stress had already been investigated in England by Hodgkinson (see Fig. 151) and others. But the introduction of the stress method of design into bridge engineering led to numerous inquiries and investigations on the permissible stress for moving loads; for instance, Fairbairn carried out, in 1860–62, a remarkable experiment with a 20-foot plate girder, subjected to a pulsating load applied at the rate of 8 cycles per minute.

The experiment showed that the metal would withstand over 3 million cycles of repetitive stress, attaining 6.25 tons per square inch; notwithstanding which, the limit fixed by the Board of Trade remained unaltered.

The significant point of the difference between the ultimate-load and the working-stress methods, is as follows: had iron been elastic under the full range of stress from zero up to the breaking-point, the two methods would have given the same result, because then the ratio of the ultimate load to the working load would have been the same as that of the breaking stress to the working stress. However, as the limit of elasticity for structural iron and steel is far below the breaking point, it follows that when rupture occurs, the proportionality of load to stress ceases to hold good, and the coefficient of safety calculated from

$$\frac{\text{ultimate load}}{\text{working load}}$$

differs substantially from the factor

$$\frac{\text{ultimate stress}}{\text{working stress}}.$$

This point deserves attention—at the present time probably more than ever before—because after a century of almost universal use, the working-stress design criterion may have, possibly, outlived its time. Whether this is indeed so remains, of course, to be seen, but the fact of the matter is that on both sides of the Atlantic, numerous suggestions have lately been put forward[1] tending to include the plastic part of the stress-strain curve—i.e. that part of it lying beyond the conventional elastic limit—into the range of conditions employed by the engineer for design purposes; and thus, to base the safety concept on the ultimate load, rather than on the working stress.

It is to be noted that the everyday practice of the drawing office, insofar as structural steel is concerned, has not yet been affected by these new tendencies, for the bulk of the structural design still follows (and will follow for some little time to come) the classical theory. But the idea of porportioning metal-work for the plastic range has already gained a strong foothold in the mind of the progressive designer, and has crystallized into a number of workable solutions, capable of being used for the quantitative design of structures; e.g. Mr. A. Hrennikoff's "Theory of Inelastic Bending with reference to Limit Design", in *Trans. Am. Soc. C.E.*, Vol. 113, 1948, p. 213.

For the time being, however, the classical design method, based on the working-stress criterion, still holds the field. The argument of this method—in its historic perspective—must,

---

[1] *Cf.* Van den Broek, *Theory of Limit Design* (J. Wiley & Sons, New York, 1948). W. Prager, "Recent Developments in the Mathematical Theory of Plasticity", *J. Appl. Physics*, March 1949. S. Feinberg, "The Principle of Limiting Stress", *Applied Mathematics and Mechanics*, 1948, Vol. 12, pp. 63–68. D. Drucker, "Plasticity of Metals—Mathematical Theory and Structural Applications", *Proc. Am. Soc. C.E.*, Vol. 76 Separate 27, August 1950, and others.

therefore, be fully digested, before any attempt is made to examine more closely the recently proposed departures from the consecrated design solution it embodies.

The first important wrought-iron bridges on the Continent were built by Lentze, over the Weichsel at Dirschau, and over the Nogat at Marienburg (see Fig. 152) in 1850–57. Lentze's original intention had been to erect a complete span in the workshops, and to test it to breaking point, in order to ascertain its ultimate resistance; but in view of the successful completion of the "Britannia" bridge in Wales he decided to dispense with this costly experiment and the final design of the Dirschau bridge was therefore adjusted according to information received about the "Britannia" bridge.[1]

The evolution of the stress criterion in Europe starts from the theoretical investigations of Schwedler and Culmann (1851 and 1852). Prior to this, the capacity of the web members of a bridge to resist shear was not clearly understood, and the designers contented themselves with calculating the area of the chord members only, according to a simplified version of Navier's bending theory.

In about 1852, Werder designed and built for the State Railways of the Government of Saxony, a 100-ton testing-machine which was arranged for testing complete, built-up bridge members. It was used for determining the resistance of the elements of the framework of the iron bridge built by Gerber over the Isar, near Gross-Hesselohe, in 1857–58.

The following instances will give a general idea of the stress limits adopted on the European continent during this period: Rebhan, in his treatise on timber and iron structures published in 1856, recommended 670 kg/sq cm for iron, in tension and in compression; Schwedler suggested 684 kg/sq cm, in 1862; the Union of Austrian Architects and Engineers adopted, in 1866, a limiting stress of 790 kg/sq cm.

The aggravating influence of moving loads was well realized at this period, but was not covered numerically by any specific safety factor, so that the same permissible stress was adopted for all the members of the bridge, regardless of any consideration of impact. Broadly speaking, up to the end of the seventh decade of the century, no account was taken quantitatively of the fact that the influence of moving loads was different for different constituent parts of the anatomy of an iron bridge.

An exception to this rule may be found in the designs of one of the greatest engineers of this period, namely, Gerber. As early as 1857 he devised the formula:

$$f = 1{,}600 \, \frac{E-V}{E+3V}$$

where f was the permissible stress in kg/sq cm,

E was the total stress due to structural weight,

V was the total stress due to moving loads.

According to this equation the permissible limit was a function of the ratio of permanent to variable stresses.

Gerber's suggestions were probably too advanced for the average engineer of the period, and were not accepted by the profession. The principle which involved defining the permissible limit for stress by means of a formula became generally recognized much later, namely, only after the publication of the experiments of Wöhler, carried out in the period 1859–70.

---

[1] The Dirschau bridge, like the "Britannia" bridge, consisted of a set of continuous girders. The maximum effective spans were 141.73 metres in the "Britannia" bridge, and 130.88 metres in the Dirschau bridge.

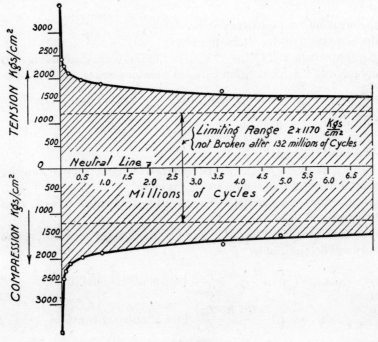

Fig. 153.

4. Wöhler's Tests

Wöhler's results may be summarized as follows:

(1) The resistance of steel and iron under variable stress depends on the *range of variation* of the stress to a greater extent than on its *absolute intensity*.

(2) Even *very low reversed* stresses are capable of causing fracture, provided the *number of cycles* is high enough.

The effect of the number of cycles on the critical range of stresses causing fracture, is illustrated in Fig. 153.

The diagram is based on a series of tests carried out by the Phoenix Company with axle-iron.

The material was subjected successively to equal tension and compression loads. The abscissae represent the number of cycles causing rupture. The ordinates are the intensities of stresses at which fracture actually takes place. The vertical distance between the two lines therefore shows the critical range.

The diagram appears to indicate that the breaking range tends to a limit, for which the critical number of cycles is infinite. This limit is called the "limiting range" of stresses. For the case illustrated, it is equal to about 2,340 kg/sq cm, which corresponds to an absolute intensity of about 1,170 kg/sq cm. It is interesting to observe that under a steady load, the ultimate resistance of the same material was about 3,600 kg/sq cm, that is to say, almost three times as much as under reversed stress.

The analysis of Wöhler's experiments led Gerber to suggest the following equation

$$(f_{max}+f_{min})^2+\frac{2V^2}{B}(f_{max}-f_{min})=4V^2,$$

where V was the breaking limit for steady stress
B was the breaking limit for alternating stress.

Plotting f_{min} as abscissae and $(f_{max}-f_{min})$ as ordinates, we get, according to this equation, a parabola, representing the relationship between breaking range and critical intensity. The average value of the ratio $\dfrac{U}{B}$ according to Wöhler's and Bauschinger's experiments, is 0.3 for iron, and 0.25 for steel.

5. Launhardt-Weyrauch Formula

Another formula based on Wöhler's experiments was developed by Launhardt in 1873. It covered the effect of repeated (but not reversed) stresses.

Launhardt's investigations were later generalized to apply to all types of variable stresses, by Professor Weyrauch, in 1876.

According to the Launhardt-Weyrauch formula the relation between breaking range and breaking intensity was determined by

$$f_{breaking}=C\left(1+\frac{U-C}{C}\times\frac{f_{min}}{f_{max}}\right) \text{ for unidirectional stress.}$$

$$f_{breaking}=C\left(1-\frac{C-B}{C}\times\frac{f_{min}}{f_{max}}\right) \text{ for alternating stress.}$$

C represents here the breaking intensity when the stress varies from zero to a maximum.

Including a factor of safety to cover the dynamic effect and other unforeseen causes of stress, Weyrauch arrived at the following general formula for the permissible working limit:

For unidirectional stresses
$$\begin{cases} f=700\left(1+\tfrac{1}{2}\times\dfrac{f_{min}}{f_{max}}\right) \text{ kg/sq cm for wrought iron.} \\ f=1{,}100\left(1+\tfrac{9}{11}\times\dfrac{f_{min}}{f_{max}}\right) \text{ kg/sq cm for steel.} \end{cases}$$

For alternating stresses
$$\begin{cases} f=700\left(1-\tfrac{1}{2}\times\dfrac{f_{min}}{f_{max}}\right) \text{ kg/sq cm for wrought iron.} \\ f=1{,}100\left(1-\tfrac{5}{11}\times\dfrac{f_{min}}{f_{max}}\right) \text{ kg/sq cm for steel.} \end{cases}$$

The Launhardt-Weyrauch formula has been extensively used in the late nineteenth century, and appeared with some slight modifications in many standard specifications of the late nineteenth and early twentieth centuries. For instance, in the Official Circular of the Swiss Union, dated 1892, the permissible stress for structural steel was

$$f_{max}=800+250\frac{f_{min}}{f_{max}}.$$

In this formula the tensile stress is considered positive, and the compressive stress negative.[1]

---

[1] As will be shown later on, the Swiss Union specification of 1913 is another form of the same equation.

Notwithstanding the recognized academical value of Wöhler's experiments, many objections were subsequently raised against the direct application of his results to practical bridge design. Mohr and Landsberg may be mentioned among those who criticized Weyrauch's formula.

From among the various aspects of Mohr's criticism, the following objections call for particular mention:

1. A general formula giving the permissible stress for any bridge member should include, in addition to the calculated theoretical stress, many other factors affecting the safety coefficient, as for instance: effect of shocks produced by the rolling loads, length of the period of application of the loads, precision (or otherwise) of the calculation, relative importance (for the safety of the entire work) of the particular member under consideration. Since all these factors cannot possibly be included in one and the same practically applicable equation, Mohr objected in principle to the use of a general formula universally applicable to all cases of bridge design.

2. The Launhardt-Weyrauch formula was based, to a great extent, on experiments with test-pieces stressed beyond the elastic limit; whereas the calculated stresses were actually derived from the assumption that the metal was ideally elastic.

Further evidence bearing on the same point was supplied by Bauschinger. The result of his comprehensive experiments was that for stresses of the same denomination, the test-pieces were still unbroken after many millions of cycles, *provided the variable stress remained within the limits of elasticity*.

The point is of obvious importance, because, according to the classical, "elastic" design principle, in order to *prevent permanent deformation*, the practical stress must never reach the elastic limit, and therefore for the case of unidirectional stress, the permissible intensity is fixed in relation to the elastic limit. Until this limit is reached there is hence no possibility of failure (under unidirectional stress) consequent on the aggravating influence of stress variation. Accordingly, in some specifications the Launhardt-Weyrauch formula was applied only in the cases in which the stress was reversed, but not otherwise. For instance, the specification of the Egyptian State Railways of 1927 stipulated that the members of the main trusses subjected alternately to tension and compression, were to be designed in such a way that the permissible stress should not exceed

$$f_{max} = 850 - 450 \frac{f_{min}}{f_{max}},$$

but the limit for unidirectional stress was a constant.

SECTION II. LATER PERIOD

1. Methods of Dealing with Impact Effect

Another line of thought relating to the same problem was inaugurated with the introduction, at about the end of the ninth decade of the last century, of a variable factor of safety depending on the estimated magnitude of the dynamic effect. With regard to this factor two different methods have been followed.

In the first method the impact stress was not calculated numerically, but a series of graduated limits were adopted for the static stress; these limits being fixed (for the various members of the bridge) in such a manner that the greater the probable dynamic effect, the smaller the permissible static limit.

In the second method a single standard limit was adopted for all the members of the metal structure of the bridge, but the static stress due to the live load (alone) was multiplied by a certain variable coefficient which represented quantitatively the estimated intensity of the dynamic effect for that particular member.

The well-known specification prepared by Sir Benjamin Baker in 1892, and extensively used in Great Britain for quite a long time, was a very successful example of the first type; the permissible stresses were given as follows:

Span of bridge	Permissible stress, in tons/sq in	in kg/sq cm
Under 20 feet	$4\frac{1}{2}$	708
From 20 feet to 25 feet	$4\frac{3}{4}$	747
,, 25 ,, ,, 30 ,,	5	787
,, 30 ,, ,, 50 ,,	$5\frac{1}{4}$	826
Above 50 feet	$5\frac{1}{2}$	865

These limits were applicable to rail bearers, cross-girders and main girders (if the latter were of plate-girder construction). For lattice girders the limits were the following:

Span of bridge	Permissible working stresses			
	Top and bottom chords (booms)		Web members, i.e. verticals and diagonals	
	In tons per sq in	In kg per sq cm	In tons per sq in	In kg per sq cm
80 feet to 160 feet	$5\frac{1}{2}$	865	$4\frac{1}{2}$ to $5\frac{1}{2}$	708 to 865
160 ,, ,, 200 ,,	$5\frac{3}{4}$	905	$4\frac{1}{2}$ to $5\frac{1}{2}$	708 to 865
200 ,, ,, 400 ,,	6 to 7	944 to 1,101	$4\frac{1}{2}$ to 7	708 to 1,101

The $4\frac{1}{2}$-ton stress for diagonals was intended to be applied to the central portion of the span only, and included the counterbracings also. On the other hand, the higher limits were applicable to similar members at either end of the span, where the variations of stresses were less pronounced.

The permissible stress for wind bracings and floor suspenders for all spans was $8\frac{1}{2}$ tons/sq in (1,337 kg/sq cm), and $2\frac{1}{2}$ tons/sq in (393 kg/sq cm), respectively.

The reason of the variation of the stress limits in this specification included both the effect of impact and the destructive action of stress variations.

The permissible stresses adopted in several bridges of the Egyptian State Railways represented a simplified application of the same principle; namely, the permissible tension stress was fixed as follows:

450 $\frac{kg}{cm^2}$ for rail bearers and plate girders directly supporting the sleepers

550 ,, for cross-girders

850 ,, for main girders.

The objective of the varying scales of stresses was confined, in this case, to the impact

SPECIFICATIONS FOR STEEL BRIDGES

effect alone, since the "fatigue" was accounted for by the Launhardt formula, which was used in this specification exclusively for reversed stresses.

It will be seen that the permissible limits adopted in this case were considerably lower than the standards of Sir Benjamin Baker's earlier specification. This was particularly true for cross-girders and rail bearers. The very high coefficient of safety thus introduced was supposed to provide, in the Egyptian specification, for the eventual development of future traffic, resulting in a potential increase of the weight of locomotives.

It seems however, to be more logical to allow for such future increase by adopting a conventional train, which is heavier than those circulating on the line at the time of the publication of the specification; and therefore, in preparing, in 1926, the first general (i.e. standard) specification for the Bridges Service of the Egyptian State Railways, the author raised the maximum permissible tensile stress to 1,100 kg/cm^2 (instead of 850 kg/cm^2). Six years later, in the second edition of this specification, the limit was raised to 1,400 kg/cm^2, disregarding secondary stresses, and to 1,600 kg/cm^2, including such stresses. Now this limit is raised to 2,100 kg/cm^2 for special steel (French, S.T. 52).

Another specification which may, in a way, be considered as belonging to the same group, was published in Switzerland in 1913. The effect of the dynamic stress was accounted for in this case by variable coefficients in the Launhardt-Weyrauch formula; namely, the permissible stress was given as follows:

For railway bridges $\quad f_{max} = 900 + 200 \dfrac{f_{min}}{f_{max}}$

For road bridges $\quad f_{max} = 1,000 + 200 \dfrac{f_{min}}{f_{max}}$

For buildings and roofs $f_{max} = 1,100 + 200 \dfrac{f_{min}}{f_{max}}$

Apart from these, one might say "historical" instances, the graduated-specification concept is also frequently evoked, although not necessarily in the same light, in various publications dealing with the same problem; for instance, Messrs. R.V. Ivy, T.V. Lin, Stewart Mitchell, N.C. Roab, V.J. Richey and C.F. Scheffey produce, in a paper on "Live Loading for Long-Span Highway Bridges" in the *Transactions of the American Society of Civil Engineers*, Vol. 119, 1954, p. 992, the following table, which speaks for itself:

Load Ratios, Stresses, and Safety Factors for Various Bridge Components

Member	Ratio of live load to dead load		Unit stress, in pounds per square inch		Safety factor	
	300-*ft* span	1,000-*ft* span	300-*ft* span	1,000-*ft* span	300-*ft* span	1,000-*ft* span
Slab	28.0:1	28.0:1	18,200	18,200	1.8	1.8
Stringers	4.0:1	3.0:1	19,200	19,500	1.7	1.7
Floorbeams	2.0:1	1.0:1	20,000	21,000	1.7	1.6
Verticals	1.1:1	0.7:1	20,800	21,500	1.6	1.5
Hangers	0.8:1	0.6:1	21,400	21,800	1.5	1.5
Web-members	0.6:1	0.2:1	21,800	23,000	1.5	1.4
Chords	0.3:1	0.2:1	22,600	23,000	1.5	1.4

The second method, which includes an explicit, *numerical* estimate of the conventional dynamic stress, is usually referred to, at present, as the "impact" method.

We have already stated (p. 170) that it was first suggested by Gerber (in 1857), but the more recent type of the impact formula is due to an American engineer, C.C. Schneider, who introduced it, in 1887, into the specification of the Pencoyd Bridge Company. Schneider's formula was therefore generally known as the "Pencoyd" formula. It was as follows:

$$f_i = f_e \frac{300}{L+300},$$

where f_i was the impact stress, to be added to the static live stress, f_e was the calculated static live stress, and L was the length of the loaded distance, in feet, which produces the maximum stress.

In metric units the same formula reads

$$f_i = f_e \frac{91.6}{L+91.6} \text{ or, in round figures } f_e \frac{90}{L+90}$$

where L is expressed in metres.

The permissible limit for the sum of the dead, live and impact stresses adopted in the Pencoyd Specification was as follows:

$$15{,}000 \text{ lb/sq in } \left(\text{or, } 1{,}055 \, \frac{\text{kg}}{\text{cm}^2}\right) \text{ for soft steel}$$

$$17{,}000 \text{ lb/sq in } \left(\text{or, } 1{,}200 \, \frac{\text{kg}}{\text{cm}^2}\right) \text{ for medium steel.}$$

The same limit was also used for members subject to stresses due to wind pressure, centrifugal force, and momentum of train.

The main objection against this formula centres on its rather arbitrary origin; on the other hand, the proof of its reliability lies in its extensive use (in America) during about half a century (though the same argument applies also to many other specifications which *do not include* the calculation of the impact stresses at all).

Various attempts were made to develop an impact formula on a reliable theoretical basis; but it must be realized that many problems must be solved before such a presumably flawless formula can be computed, owing to the many uncertain factors affecting the vibrations of a bridge member. For instance, both from the theoretical and from the experimental viewpoints, the magnitude of the impact stress in a bridge depends on the coincidence of its natural oscillation rate, with the periodicity of the impulses, from the unbalanced elements of the locomotive.

The particular speed of the train at which these periods coincide with the frequency period of the natural oscillations vibration of the bridge is usually termed the "critical speed". It corresponds to the *maximum impact* effect. The frequency period of natural oscillation and the critical speed may be determined by calculation,[1] and some of the tests carried out on

---

[1] The frequency period of natural oscillation can be calculated from the formula

$$T = \sqrt{\frac{W+P}{P} d},$$

where T is the period of oscillation of a loaded bridge in seconds, W is the dead load per foot, assumed as uniform, P is the live load per foot, also assumed as uniform, and d is the static deflection in feet, due to the load P, as determined by direct measurements.

SPECIFICATIONS FOR STEEL BRIDGES

actual bridges show that the recorded critical speed agrees fairly well with the one so calculated. In effect, the critical speed varies from 65 miles per hour, for a 60-ft span, to 25 miles per hour, for a 300-ft span. Yet it would be highly impracticable to include such calculations in the design of bridges, because, apart from the tremendous difficulty of the calculation itself, the stress would then depend, to a great extent, on the kinetic characteristics of each type of locomotive running on the line.

Systematically planned experimental investigations on impact stresses in bridges were first started by the American Railway Engineering Association, in 1907; and since then, similar experiments were made in India by the Railway Bridge Committee, in England by the Ministry of Transport, and by many other authoritative bodies in various countries.

Investigations of this nature are carried out by means of special instruments, recording the local strains in different bridge members during the passage of a train—at different speeds. Within the elastic range, impact may be estimated from the percentage difference of strains corresponding respectively to high and low speeds of the same train (say, 60 miles per hour as against 3 miles per hour). However, the main difficulties to be faced in deriving significant conclusions from such observations, are as follows:

(1) The measured extensions are of a *very small* order of magnitude, for otherwise they could not give *local* values of stress. Consequently, the recording instrument must be highly sensitive.

(2) The individual vibrations of the movable parts of the *instrument itself* may confuse the records, at high speeds of vibrations.

2. The Fereday-Palmer Instrument

Various devices are used for carrying out such observations. To give an example of the earlier type we shall describe the method of the Fereday-Palmer photo-recording instrument,

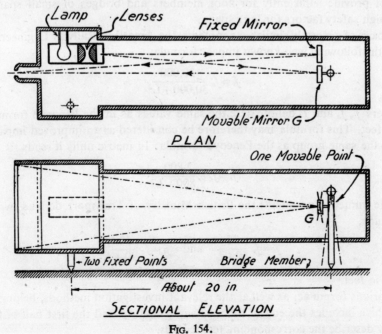

Fig. 154.

which is chosen as a typical instance of the group to which it belongs. It is schematically shown in Fig. 154. Except for certain types of troughing and rail bearers, the frequency of the instrument (about 200 to 300 cycles per second) was considerably in excess of the natural frequency of the bridge members under test, and consequently, disregarding some minor defects, the records were considered sufficiently accurate.

The instrument consists of a hollow cylinder, about 2 feet in length, provided with three hard-steel points, which constitute the contact between the apparatus and the bridge member investigated (see Fig. 154). The movable point is connected with the *movable* mirror G, which reflects the images of the slits on a photographic film. Another image of the slit is reflected by the *fixed* mirror. A disc, operated by a clock-work mechanism, periodically covers the fixed mirror (for one quarter of a second) and uncovers it again for another quarter of a second. A series of dashes is thus obtained on the film, which permits the estimation of the frequency of the vibrations of the image reflected by the movable mirror.

The records of the instrument are in the form of two lines, a—b and c—d (see Fig. 155). The strains are scaled off from the line a—b, for they are proportional to the distance x and the stresses deduced therefrom. In the classical experiments of the British Ministry of Transport, which were made with this apparatus, the recorder was calibrated to one-tenth of an inch per ton, per square inch, assuming perfect elasticity.

FIG. 155.

The most significant conclusion derived from these experiments was that the Pencoyd formula did not provide sufficiently for floor members and bridges of small spans, but gave excessively high safety factors for long spans.

On the basis of generally similar experiments, the American Railway Engineering Association devised the following well-known impact formula:

$$f_i = f_e \frac{30{,}000}{30{,}000 + L^2}.$$

The letters f_i, f_e and L denote here the same values as in the Pencoyd formula and L is expressed in feet. This formula may therefore be considered as an improved impact equation, belonging to the same group as the Pencoyd formula. In metric units it reads

$$f_i = f_e \frac{2{,}800}{2{,}800 + L^2}.$$

From the quoted experiments the British Ministry of Transport derived two alternative impact formulae,

$$f_i = f_e \frac{75}{50 + L} \quad \text{and} \quad f_i = f_e \frac{120}{90 + L}$$

where L was in feet.

These various formulae, as well as the relevant investigation methods, belong basically to the period which includes the end of the nineteenth century and the first half of this century. We shall now describe the corresponding test records.

SPECIFICATIONS FOR STEEL BRIDGES

It must be clear from the foregoing discussion that the chief difficulty in obtaining an accurate measurement of instantaneous stress intensities lay in the movable parts of the recording apparatus, for however small these parts might have been, their inertia sufficed to vitiate the records. It was therefore only natural that further developments in this branch of engineering science tended to eliminate these movable parts altogether, and this could only be achieved by using electrical appliances.

3. Electrical Measuring Instruments

In fact, two classes of electrical measuring instruments were produced—the electro-magnetic and the wire-resistance types, respectively. In the former[1] the strain gauge is attached to the member by two screw seated in specially drilled and tapped holes, whereas in the second type the gauge is cemented by means of various cementing media to the surface of the member. In both cases the gauge is placed in one arm of an electrical bridge circuit, whilst a similar, "balancing", gauge is placed in the other arm of the same circuit. In the electro-magnetic device a strain in the auscultated member produces a change in the electrical impedance of the circuit, whilst in the wire-resistance instrument, a similar strain causes a change in the electrical resistance of the wires bonded to its surface. Here, therefore, as in all earlier examples, we measure *strains* and deduce therefrom *stresses*.

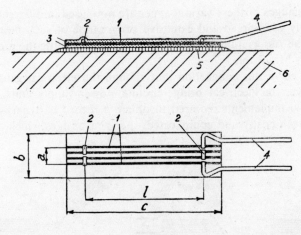

FIG. 156.

Well-known types of such measuring instruments are the famous SR-4 gauges, manufactured by Baldwin-Southwark, in America, and the Tepic gauges made by Huggenberger, Zürich, on the European continent. Both are of the wire-resistance type, and are, of course, patented.

To give an example, attention is called to Fig. 156, which shows diagrammatically a Tepic gauge.

The very thin wires (1), the electrical resistance of which is used as a measure of the strain, are connected with each other by short conductors (2), having a much larger sectional area.

---

[1] In drafting the following paragraphs free use has been made of Mr. E. J. Ruble's paper on "Impact in Railroad Bridges" published, in July 1955, as Separate No. 736 by the American Society of Civil Engineers, New York. Figs. 160 to 162 are also abstracted therefrom.

This arrangement is a specific characteristic of Huggenberger's type, which is claimed to render the instrument insensitive to transverse strains—a much more difficult problem to solve, if the more usual "multiple loop" layout is adopted (as in Fig. 157).

The outer appearance of a Tepic gauge is as shown in Fig. 158.

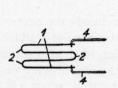

Fig. 157.

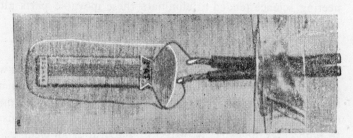

Fig. 158.

In various stress problems it is frequently required to know the simultaneous intensities of the stresses acting in different directions in the same point, or to use the more technical term, "the rosette" of stresses. This problem is solved empirically by combining several electrical strain guages, for instance as shown by Huggenberger in Fig. 159.

Electrical strain gauges of these various types are now used in different branches of engineering. A particularly widespread and extensive programme of such measurements, undertaken since 1941 by the Association of American Railroads to settle the problem of impact in bridges, is reported by Mr. Ruble.[1]

It must be realized in this connection that though the electrical gauges themselves are simple and inexpensive, the electrical bridge circuits, power supply units, oscillographs and various other electrical implements required to obtain a record of dynamic strains by means of such gauges, are both costly and complicated. For the bridges which were accessible by

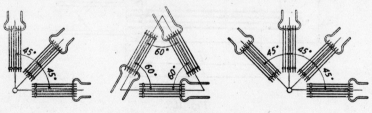

Fig. 159.

highway, private road or pasture, all this auxiliary apparatus was installed on a truck. For inaccessible bridges, special housing facilities had to be resorted to.

A typical record obtained by Mr. Ruble for a plate-girder bridge is represented in Fig. 160.

The twelve oscillograms shown on the left-hand side of the drawing were recorded under a steam locomotive running at 53.4 miles per hour. The location of the corresponding twelve gauges is given in the sketch on the right-hand side of the same chart.

The abscissae of the oscillograms represent strains, while the ordinates give the corresponding times. Every time a wheel passes over the centre of the span, a mark is placed on the

[1] See foot note, page 179.

SPECIFICATIONS FOR STEEL BRIDGES

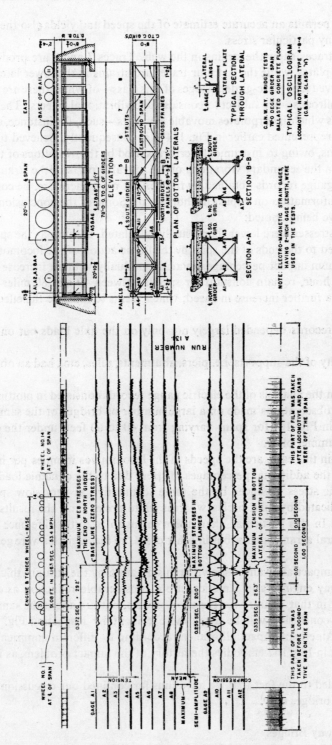

Fig. 160.—Typical oscillogram showing stresses recorded in railroad girder bridge.

oscillogram, which permits an accurate estimate of the speed and yields also the position of the engine at time of any particular stress.

The upper six traces show the strains in the web. Traces 7 and 8 are produced by strains in the lower flange plates, while the last four traces are strains in the lower lateral bracings.

It will be observed that plate-girder bridges of this class—of which there are some 36,000 on the United States railroads—are subject to considerable vibrational effects. The type of strain-recording apparatus which incorporates movable elements—such, for instance, as the Fereday-Palmer instrument represented earlier in Fig. 154—is, consequently, believed to be unsuitable for use on such spans, owing to instrumental errors caused by the vibrations of the entire metal work, as reflected in the secondary vibrations of the movable parts of the strain recorder.

The electrical gauge records obtained by the A.A.R. may, therefore, be considered as the first, fully reliable information on impact in short spans. Some of the conclusions derived from these records deserve being quoted:

(1) The static stresses were lower than usually calculated, because as the span deflects, the axle loads are carried to the ends of the span by the frame action of the locomotive.

(2) The maximum impact percentages increase gradually with an increase in speed up to about 40 miles per hour, remain constant for speeds between 40 and 80 miles per hour, and then decrease with a further increase in speed; which seems to indicate the effect of synchronous speeds.

(3) The stress records depended largely not only on the axle loads but on their arrangement also.

(4) The elasticity of the supports, i.e. piers, abutments, piles, etc., had an obvious influence on the results.

The next step in the analysis of the electric gauge records consisted in plotting in one single chart the results of observations made on a large number of bridges of the same category; for instance, as shown in Fig. 161 for spans varying from 40 to 60 feet, under the effect of steam locomotives with hammer blow.

The abscissae, in this case, are the speeds of the locomotives in miles per hour, while the ordinates represent the additional percentages of stress due to the dynamic load, as compared with the basic, static stress produced by the same locomotive at very slow speed. Ballasted deck bridges are indicated by solid circles, whereas white circles show the results obtained with open deck bridges. In this particular chart there seems to be little difference between these two types of structural arrangements, but in other cases the black circles were generally located lower.

The proposed impact allowances shown in this diagram by the straight thick lines, and already used by railway engineers in America, are a considerable reduction as compared with those formerly used in that country for the short girders, but are about the same for the longer spans. This last conclusion will be made more evident by inspecting Fig. 162, in which Ruble has plotted American impact allowances according to different empirical formulae. In that sense, the chart in Fig. 162 embodies the history of the impact problem, as it developed in the United States.

Attention is called to the fact that we have thus far confined our discussion of the impact problem to railway bridges only.

4. Impact for Highway Bridges

The question of impact percentage for highway bridges is possibly of less importance. Some

SPECIFICATIONS FOR STEEL BRIDGES

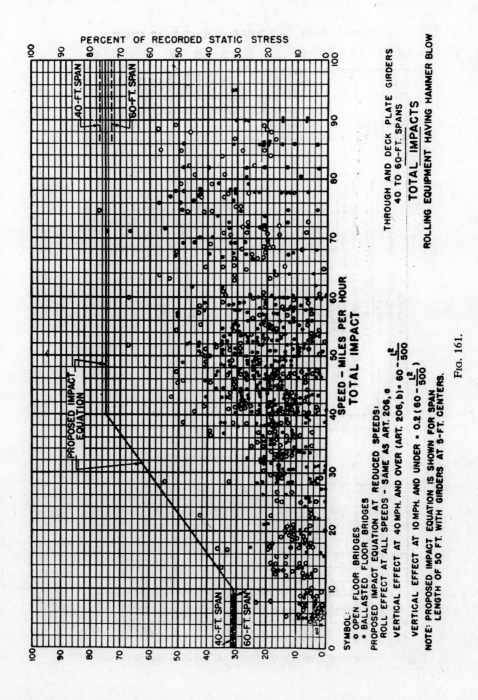

Fig. 161.

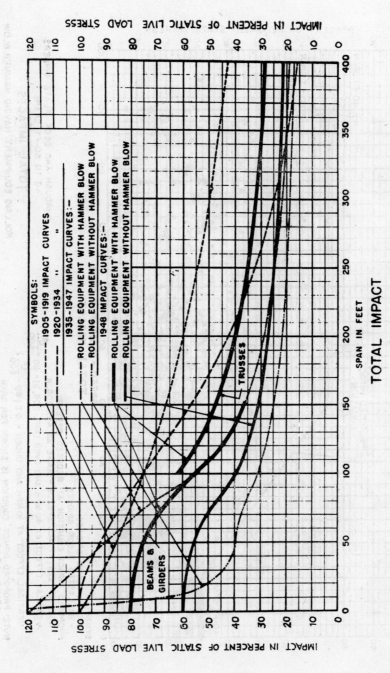

Fig. 162.—Historical development of impact equations in America.

designers assume in this case a constant impact coefficient having a value from 0.25 to 0.5, while others prefer to neglect the impact for road bridges altogether. This latter procedure, however, does not seem to be fully justifiable, because experiments show that highway bridges as well as railway bridges, are subject to dynamic stresses, which are by no means negligible. Such experiments were, for instance, made by the Iowa Experimental Station. Though no specific formula was then arrived at, the published records indicated an impact effect of about 15 to 25 per cent, for a perfectly smooth road surface. If small observations, such as 2- to 4-inch planks, were placed on the road under the wheels of the car, the impact increased up to 50 per cent, and even more.

A contemporary well-known American bridge designer, J.A.L. Waddell, employed, for highway bridges, the following impact formula:

$$f_i = f_e \frac{100}{\frac{WL}{20}+200}$$

in which f_i, f_e and L had the same meaning as in the preceding equations, while W was the sum of the widths of the roadway and footwalks. Both L and W were expressed in feet.

In Russia, the following formula was adopted at a somewhat earlier period, for the highway bridges of St. Petersburg:

$$f_i = f_e \frac{75}{100+L}$$

where L was expressed in metres; the permissible stress for the sum of static and impact stresses was 1,200 kg/sq cm.

According to the British Standard Specification of 1937, the impact effect for road bridges was as follows:

$f_i = f_e (0.75 - 0.002L)$ with a maximum $f_i = 0.60 f_e$ for one line of traffic, and

$f_i = f_e (0.65 - 0.002L)$ with a maximum $f_i = 0.50 f_e$ for two or three lines of traffic,

where L = effective span in feet.

5. Impact Effect Dealt with as a Function of the Speed of Vehicles

Logically, the impact effect must depend on the speed of the vehicles. In fact, the *Final Report of the 2nd Congress for Bridge and Structural Engineers*,[1] 1929, contains the following formula framing this principle:

$$\frac{f_i}{f_e} = \frac{1.5}{1+\frac{V}{3v}},$$

in which v is the speed of a single motor vehicle, expressed in miles per hour, whilst V is a parameter, namely: a conventional speed of 10 miles per hour. This formula is obtained empirically from observations, as a curve of best fit.

Professor Freudenthal generalizes this equation by assuming several vehicles, and obtains[2]

$$\frac{f_i}{f_e} = \frac{1.5}{\left(1+\frac{V}{3v}\right)\sqrt{n}},$$

[1] A.H. Fuller, "Impact in Highway Bridges", p. 56.
[2] A.M. Freudenthal, "The Safety of Structures", *Trans. Am. Soc. C.E.*, Vol. 112, 1947, p. 146.

in which n is the number of vehicles on the span. This formula is obtained from the former one, assuming that the probability of the simultaneous occurrence of the dynamic increment of several vehicles is governed by chance.

Note however, that according to modern views on the subject, n, the number of vehicles on the span, is again a function of their velocity, for according to investigations of the U.S. Bureau of Public Roads, the spacing between the centres of the vehicles is capable of being transcribed by the formula

$$l = \frac{L}{1 - 0.15 v/V} (1 \pm 0.20),$$

in which L is 40 feet for trucks and 30 feet for passenger cars, whilst V is, as before, 10 miles per hour. The formula is valid for speeds, v, up to 60 miles per hour.

Various other formulae representing the same effect have also been suggested. For instance, in the paper already quoted earlier (see p. 175) R.V. Ivy and his associates use the formula

$$l = K_1 v^2 + K_2 v + a + b,$$

in which l is, as before, the spacing of the vehicles
    K_1 is the vehicle constant $= 0.5$
    K_2 is the driver reaction constant $= 1.5$
    a is the length of the vehicle
    b is the minimum distance between vehicles, usually 8 feet.

Whichever formula is used, it is obvious that a large impact factor for highway bridges presupposes a high velocity, and therefore a smaller number of cars on the span, i.e. a lighter load. This correlation however does not apply to railway bridges, because in this case the spacing of the wheel loads is independent of the velocity. This postulate reflects the basic difference in the philosophies of the impact factor for railway and roadway bridges respectively.

Much discussion has at some time, taken place in connection with the very heavy concentrated loads prescribed for the design of some highway bridges in Europe. For instance, some British bridges were designed for a 180-ton truck, crossing the bridge may-be once or twice during its lifetime. Note that in this case the prescribed limit of tensile stress was slightly higher than the yield-point—a contingency deserving attention both in respect to its intrinsic significance and also as symptomatic of the effect of the plastic theories on the designer's credo (some of which have already been quoted on p. 144, whilst others will be mentioned later).

However that may be, vehicles of such weight will, of course, be assumed to crawl rather than to move, and no impact percentage will therefore be added over and above their specified weights.

SECTION III. SPECIAL ALLOY STEELS

1. Different Kinds of Steels Used in Parallel in the Same Bridge

One of the features of modern structural metal-work, characteristic of its up-to-date design tendencies, is the use of special high-resistance steels. A new principle is thus introduced into the specifications and design methods; viz. the individual members of the steel structure are adjusted to the stress they carry not only in respect to *their dimensions*, but also in varying the *class of material* from which they are manufactured. The evolution of this general idea may possibly be in its infancy, for usually there are a few different alloys only, which are incorporated in one and the same structure; but there is a wide field for possible future developments, until graduation of materials will be as easily done as that of the sectional areas of the members.

The use of special steel was first motivated by considerations of economy in the design of long-span bridges such as those lately built in the U.S.A.; for in spite of being more expensive, this material carries a higher stress, and consequently allows a reduction of the structural weight of the span. This is a predominant consideration for long spans in which the dead-load effect is the significant factor, being far more important than the action of the live load. For instance, in the George Washington bridge the ratio of dead load to live load is as 5 to 1; in the Golden Gate bridge it is as 5.25 to 1, and so on.

The overall saving consequent on the use of high-strength materials in such bridges, balances—and exceeds—the extra expenditure on the individual members in which these materials are incorporated.

2. Examples of Bridges in which Special Alloy Steel was Used

It appears that special nickel-steel was first used in the construction of the Queensboro' bridge, in 1903. The tension-chord load in this bridge attained 19,000 kips (thousand pounds). With medium carbon-steel (permissible stress 20 kips/sq in) this would have required about 950 square inches of sectional area. As built, the nickel-steel area was only 680 square inches.

The next application of nickel-steel was in the stiffening trusses of the Manhattan bridge. The relevant specification (1906) called for the following physical requirements:

Ultimate strength	85 to 95 kips/sq in
Elastic limit	55 ,, ,, ,,
Percentage elongation: $\dfrac{1{,}600}{\text{ultimate strength}}$	
Percentage of reduction of area	40 (minimum)

It may be noted that recent specifications for this kind of steel differ from that of the Manhattan bridge only by raising the ultimate strength from 90 to 115 kips/sq in and lowering the percentage of reduction of area to 30 per cent.

The new Quebec bridge, which was built after the collapse of the earlier structure, offers another instance where nickel-steel was successfully employed. In the report, published in 1919, the Board stated: "Nickel-steel was used wherever it would effect a saving in weight of structure as a whole, and, by this means, notwithstanding its higher price, be a factor in reducing total cost."[1]

The Delaware River bridge was the first instance in which a differentiation was made between the chords and web-members, by using different grades of steel. The stiffening girders of a suspension bridge are generally subject to relatively low shear and their web-members are consequently lighter than in the usual type of truss. They were therefore made in this case of silicon-steel, whilst the chords were in nickel-steel.

The San Francisco-Oakland bridge is a still more recent example of high-strength steel application. This was used here, chiefly in the main cantilever truss and in the anchor plates in the central pier of the suspended portion.

---

[1] Quoted from "Evolution of High-strength Steels", by Moisseiff, *Trans. Am. Soc. C.E.*, 1937, p. 1343.

The specifications for the various classes of metal used in this bridge were as follows:

	Unit stress, in kips per square inch		
	Ultimate strength	Yield point	Working stress
Medium carbon-steel	62	37	28
Silicon-steel	80	45	28
Nickel-steel	85	55	34
Heat treated eye-bars	80	50	34
Cable wire	220	150	82

As this bridge marks the beginning of a new era in important bridge design, the following table, portraying as it does the "march of time" as reflected in the use of different metals, may be of some interest:[1]

Comparison of Materials Used in Large Cantilever Bridges

Bridge	Span, feet	Weight, tons	Percentage of steel			
			Nickel	Silicon	Heat treated	Carbon
San Francisco-Oakland Bay	1,400	20,900	16	40	10	34
Firth of Forth	1,700	30,000	—	—	—	100
Quebec	1,800	62,900	27	—	—	73

On the other hand the Bayonne bridge (1927), although it is of a much smaller size and span, is of particular interest with reference to the problem we are here discussing, for it exemplifies a rational combination of *various* types of steel.

Where high stress-concentrations occur, special alloy steels must, logically, be used; but if stiffness is the predominant factor, less-resistant material may be reasonably employed, with a consequent increase in the section and weight of the member.

Conforming with this general design-principle, the material in the Bayonne bridge is graduated as follows: the main material of the lower chord (which carries the dead weight and part of the live load) is high-strength alloy steel throughout. The top chord is silicon-steel, except the three end panels which are of carbon-steel. Web-members are of carbon-steel, except where high erection stress calls for higher strength.

The silicon-steel herein referred to, is above carbon-steel in strength, but is less resistant than nickel-steel or other special steels.

3. Permissible Stress Intensities for Special Alloy Steels

The following permissible stresses are adopted by the American Railway Engineering Association for these three grades of steel:

Carbon-steel	18	kips/sq in or	1,263 kg/cm²
Silicon-steel	24	,, ,, ,,	1,690 ,, ,,
Nickel-steel	30	,, ,, ,,	2,110 ,, ,,

[1] "Actual Application of Special Structural Steel", by Beard, *loc. cit.*, p. 1308.

Silicon-steel was first applied in large quantities of weight in the construction of the steamship *Mauretania* (1907), and has been successfully used in America since about 1915 (bridge over the river Ohio, at Metropolis).

4. Various Considerations Affecting the Use of Special Alloy Steels

After the First World War, German mills adopted a similar grade of steel, containing 1.0 to 1.33 per cent of silicon.

The result was rather unsatisfactory and the attempt was therefore abandoned. This is believed to have been due to a too large percentage of silicon. The American material described as "silicon-steel" contains much less silicon, but includes a certain percentage of manganese.

There are, of course, various limitations (both structural and economical) to the use of high-strength steels. Rivets, for instance, are not as easily fabricated in these materials as rolled sections. Silicon-steel is generally believed to be a non-forging metal and cannot, therefore, be used in rivets, whilst nickel-steel rivets are difficult to drive; with the result that manganese-steel is generally preferred for making high-strength rivets, though carbon-steel rivets still provide the most popular method of joint splicing.

The low cost of petrol in the United States has stimulated the development of motor-driven traffic in that country to such an extent that toll-bridges, which in the earlier years of this century were given up as unprofitable, became again a financially sound proposition. This, in turn, led to the conception and realization of unprecedented spans, which called eventually for high-strength alloys.

5. Special Alloy Steel in the Bridges of Chelsea and Howrah. Question of Rivets

America was, therefore, *the* country where the new types of high-grade steel found their widest application; which does not mean, however, that the advantages of high-tensile steel were completely ignored in other countries. To illustrate and emphasize the point, we might quote, for instance, the reconstruction of the Chelsea bridge in England (completed 1937), and the Howrah bridge in India (completed 1943).

In the former case,[1] which was a self-anchored suspension bridge with a 352-foot central span and two 173-foot lateral spans, manganese-copper-steel was used for the $105\frac{1}{2}$-inch-wide webs, and a manganese-chromium-copper-steel for the remainder of the stiffening girders, which were in this case rather heavy, since they performed the double duty of stiffening the bridge and carrying the horizontal thrust of the "chains".

The specified and the actual physical properties of the alloy steels were as shown in the table overleaf.

It may be rather interesting to mention that in this structure the rivets were made of manganese alloy steel, of the following composition:

Carbon	0.16 to 0.22 per cent
Silicon	0.10 ,, 0.20 ,, ,,
Sulphur	0.05 per cent max.
Phosphorus	0.05 ,, ,, ,,
Copper	0.29 ,, 0.39 per cent
Manganese	0.75 ,, 0.85 ,, ,,
Chromium	0.29 ,, 0.39 ,, ,,

[1] *J. Inst. C.E.*, Jan. 1938, p. 424.

Alloy steel characteristics			Yield stress	Ultimate tensile strength	Elongation	
					Longitudinal	Cross-wise
			Tons per square inch		Per cent	
Specified			23 (*min*)	37 to 42	18 (*min*)	18 (*min*)
Actual	"Ducol" Colvilles Ltd.	Range Average	23.0 to 27.9 24.84	39.2 to 46.0 41.46	18.0 to 26.0 25.5	18.0 to 24 20.98
	"Atlantes" Cargo Fleet Iron Co.	Range Average	23.3 to 28.0 25.02	37.1 to 41.6 39.21	18.0 to 25.0 21.07	18.0 to 28.0 21.04

The tensile strengths of a rivet rod and of a driven rivet were respectively 30.9 to 34.5 and 37.1 to 46.6 tons per square inch, while the shear strength of a finished rivet was 26.2 to 31.3 tons per square inch.

The specification for the high-tensile steel of the Howrah bridge (cantilever bridge with 1,500-foot central span, between Howrah and Calcutta, over the Hooghly) was largely based[1] on the experience gained on the Chelsea bridge. Here again we find plain manganese-steel and manganese-chromium-alloy steel. In regard to chemical composition and physical properties they approach closely to those used earlier in the Chelsea bridge. In particular, the inclusion of 0.3 per cent of copper was considered essential, as smaller percentages were believed to be of little value for giving resistance against corrosion.

Extensive tests showed that in order to develop and preserve the driving qualities of high-tensile manganese-steel rivets, the percentages of carbon and manganese must be carefully controlled (0.18 to 0.22 per cent C and 0.60 to 0.75 per cent Mn) and the tensile stress of the material must be kept within 30 to 35 tons per square inch.

For instance, raising the percentages to 0.24 per cent C and 0.9 per cent Mn, produced a steel of 37.44 to 43.74 tons per square inch tensile resistance, but this was of little practical avail, for the driving qualities of the material became extremely unsatisfactory.

Periodical driving tests for manganese rivets were therefore found necessary, to ensure that the required standards were maintained.

6. Limitations for the Use of Alloy Steels

Attention is called to the fact, that all the above-mentioned instances of bridge construction exemplify practice with long spans.

It must be realized in this connection that under present-day conditions, the application of special steel in the case of relatively short spans may possibly lead to the use of awkward light sections, which are inconvenient for riveting, are subject to excessive vibration, etc. In that order of ideas, it is believed that a span of about 140 feet is the lowest limit for using silicon-steel as far as bridges are concerned.

On the other hand, in regard to the metal equipment of hydraulic structures, i.e. gates, valves, etc., the potential scope of application of high-strength steel includes also smaller spans, because the weight of a sluice-gate is not only a matter of initial investment (as happens to be the case with a bridge), but also involves a maintenance problem; for the cost of the

[1] *J. Inst. C.E.*, May 1947, p. 208.

gate-operating machinery, and its upkeep, are both functions of the weights to be lifted and of the masses to be moved.

7. Alloy Steels and Corrosion

Since, however, the steel equipment of a hydraulic work is more exposed to rust than a bridge, the point which controls this aspect of the high-grade-steel problem is, whether alloy steel is as resistant to corrosion as the usual mild carbon-steel?

In order to answer this question, we must first consider the effect which is produced on the rust-resisting property of the material by the usual non-ferrous ingredients contained in commercial steel.

The customary metalloids encountered in ordinary steel are: carbon, silicon, sulphur, phosphorus and manganese.[1]

Carbon confers to steel its essential mechanical properties (tensile resistance, elastic limit, ductility, etc.). Although it is true that as its content is gradually lowered the resistance of the material against corrosion increases, this advantage is offset by a generally lower physical standard. Within the usual percentages the influence of carbon on corrosion is, therefore, unimportant.

Manganese has often been described as a harmful agent but this is not supported by modern informed opinion. It appears that both manganese and silicon, in the quantities met with in commercial steel, may be disregarded from the corrosion viewpoint.

Phosphorus has a retarding effect on corrosion, but since it increases brittleness, its content must be kept low.

Sulphur is invariably recognized as harmful for corrosion, but its percentage is usually kept low.

It follows that, under normal circumstances, the non-ferrous admixtures in plain structural steel do not alter its non-corrosive properties; subject, of course, to the absence of manufacturing defects (such as segregation, non-uniformity, dirtiness, etc.).

Turning now to alloy steels, the most frequently used additions are chromium, copper, manganese, molybdenum, nickel, silicon and vanadium—in various combinations, with or without carbon.

Insofar as corrosion is concerned, the most promising combinations are those containing nickel, copper and chromium. The addition of nickel (up to 3.5 per cent) confers an added corrosive resistance, but the predominant consideration in this case is the high-strength characteristic of the alloy, because nickel is too expensive for use solely as a protection against corrosion; nevertheless, nickel-steel is one of the most popular low-alloy metals, owing to its excellent mechanical properties.

Copper-bearing steel is possibly the most successful anti-corrosive alloy, for even a small percentage of copper improves corrosion resistance, under atmospheric influences. Under immersion conditions the effect of copper is not as definite, but here again, a certain advantage is gained by a somewhat higher percentage of copper, which explains the use of copper-steel in sheet pilings (for instance at the Esna Barrage, on the Nile).

On the whole, there is reason to believe that, with the exception of steel containing copper, low-alloy steel may be assimilated, from the standpoint of corrosion, to ordinary steel, and

---

[1] "Corrosion in Engineering Structures" by James Aston, *Trans. Am. Soc. C.E.*, Vol. 102, 1937, pp. 1295 *et seq.*

there is therefore no objection against using the high-strength alloy material for the metal equipment of hydraulic works. It is to be remembered, however, that if the higher tensile resistance is fully utilized, the ratio of exposed to cross-sectional areas must of necessity be relatively greater in members built of alloy steels. Copper addition is therefore recommended for steel used for the metal equipment of important hydraulic works, if and when the expense on higher unit rates is justified.

In conclusion, the following table, prepared by Mr. J.C. Whetzel, gives the general percentage range of elements in various low-alloy steels[1]:

Carbon	0.10 to 0.40	Copper	0.01 to 1.40
Manganese	0.20 to 1.70	Nickel	00 to 3.5
Phosphorus	0.01 to 0.20	Chromium	00 to 1.20
Sulphur	0.05 maximum	Molybdenum	00 to 0.40
Silicon	0 to 1.0	Vanadium	00 to 0.20

SECTION IV. PHILOSOPHY OF SAFETY FACTOR

1. Generalized Concept of Safety Factor

Broadly speaking, the permissible limit of stress or load, which is to be used as basic criterion for calculating the scantlings of a structural steel-work, depends essentially on two factors: the critical limit for the material as determined mechanically in the test-room, and the coefficient of safety. The first factor may be stated either in terms of the breaking load, or of the breaking stress, or of the yield point. The breaking-load method was adopted in the earlier designs, before 1860,[2] but thereafter followed a period of about half a century, during which the working stress, calculated by the usual formulae of Hooke's and Navier's mechanics, was accepted not only as a useful, practical design parameter, but also, as the true maximum critical stress. Elastic behaviour was postulated for members and connections, and secondary stresses, such as they were supposed to be, were relegated to the realm of academic theories. The wealth of practical experience with innumerable steel structures built during this period, was almost entirely correlated with nominal, i.e. elastic, stress calculation.

Subsequent research showed that these assumptions were not realized in Nature, and that the presence of local stress-concentrations, exceeding many times the nominal, calculated working stress, could only be explained by the smoothing effect of ductility, i.e. plastic deformation. In this second period, therefore, the stress based on the elastic theory was visualized as a parameter only, but was still used as a criterion for design, the limit being fixed depending on the yield-point of the material, with a judiciously chosen factor of safety.

Viewed from this standpoint, the recent tendency to use the plastic bending theory for the design of steel structures, may well be considered as a third stage in the same evolution. Since plastic deformation is now a recognized fact explaining the safety factor incorporated in the working-stress criterion, why not broaden the approach and achieve material economy by introducing the same principle, i.e. plastic deformation, explicitly in the calculation of the stress itself? This would then allow the use of a more liberal safety factor, taking as basis the breaking load rather than the yield-point—which might possibly be exceeded. Presented in this manner these new suggestions appear, therefore, to be logical, and even self-evident, but there are, of course, many obstacles looming ahead, not the least among which is the elaborate

---

[1] *Proceedings, Am. Iron and Steel Inst.*, May 1935.
[2] See pages 167 and 168, *ante*.

mathematical work required to be performed in applying the mathematical theory of plasticity to the more complicated problems. Apart from this, many practical design problems, e.g. plastic shear concentrations at sudden changes in the shape of the profile, must be fully understood and framed by simple equations, before the plastic theory may become a generally recognized design method.

Reverting now to the classical, i.e. elastic, theory which is still universally applied in the majority of designing offices, it should be noted that for ordinary structural steel the breaking stress can be less than the elastic limit only in case the member is subject to reversed stresses. Consequently, within the limits of this classical theory, for unidirectional stress, the permissible limit depends on the yield-point alone. For reversed stresses (or for special kinds of steel with a yield-point approaching the ultimate stress) the breaking stress may occasionally be well below the yield limit, and consequently there is, in this case, sufficient reason to apply the Weyrauch formula (see p. 172).

The coefficient of safety was sometimes described as the coefficient of our ignorance. Suppose we could have precisely calculated the true stress in a bridge member: the safe limit might then have been raised to the limit of elasticity or to the ultimate load, depending on which design theory was used. But unfortunately, our lack of ideal knowledge necessitates the use of factors of safety, to compensate for the inexactitude of the calculation methods. The precise value of this factor cannot be determined exactly (for otherwise we could have included it in the calculations, and the problem would then have been automatically solved), but the factor may be guessed from laboratory research or comparison with already existing structures; or else from specifications which have repeatedly proved to give satisfactory results under analogous conditions.

2. Three Types of Error

The probable errors inherent in our classical "elastic" calculations may be classified as follows:

(1) Errors due to the unknown percentage of the dynamic effect of rolling loads.

(2) Errors due to the inexactitude or inapplicability of our elastic equations.

(3) Errors due to the possible difference between the assumed standard train, or standard car, and the actual train, or the actual car which may eventually pass over the bridge.

The first type of error is capable of being experimentally investigated and may, therefore, in the course of time, be gradually reduced by further research to a negligible value. The dynamic stresses referred to under this heading can be divided into primary, which are due to the vibration of the bridge as a whole, and secondary, produced by the individual vibrations of every member as a unit.[1]

The error under heading (2) is the difference between the actual static stress and the calculated static stress. This difference is given the title "secondary stress". The most important secondary stresses are: (1) stresses in the plane of the main truss due to the rigidity of the joints; (2) concentrations of stresses near rivet holes and cutouts of various kinds; (3) bending stresses in the members of the main frames, due to the deflection of the cross-girders, and *vice versa*; (4) non-uniform distribution of the load among rivets in tension groups, where individual end rivets take much more shear than their allotted share. Modern advanced methods of calculation allow some of these stresses to be estimated approximately, but up to

---

[1] On the average, the dynamic stresses are frequently assessed at not more than 60 per cent of the live stresses, though in particular instances much higher percentages are by no means exceptional.

the present, there is not enough experimental evidence to condemn, or to justify, the true value of the assumptions on which such calculations are frequently based.[1]

The third group of errors can evidently not be estimated and will always remain a matter for guessing.

3. Solutions Based on Mathematical Theory of Probability

Finally, to complete this brief review of the problem of the structural safety factor, attention may be drawn to the attempts at solving the problem by application of the mathematical theory of probability; the basic idea being: to replace personal intuition, which thus far had been the controlling influence in the choice of the quantitative value of the safety factor, by a scientific concept of the risk, estimated according to past experience, on the one hand, and accuracy of analytical methods as well as precision of execution, on the other. From among such attempts it will suffice to quote the well-known contribution of Professor A.G. Pugsley[2] and the concepts of Professor A.M. Freudenthal.[3]

Professor Pugsley is the only one to have produced a practical self-contained design-criterion derived in the manner described, which has actually been used in practice, but this concerned airplanes only. As seen from Fig. 163, which is abstracted from Professor Pugsley's paper,[4] the design-criterion takes in this case the form of a number of frequency curves, based on actual flight observations and correlating accelerometer readings, recording the values

$$\frac{\text{vertical acceleration}}{\text{acceleration of the force of gravity } g}$$

with the corresponding speeds. The curves yield the probability of the extreme conditions, and are indeed suitable for use, both as design-criteria for calculating the required strength of the wings, and also as guide for pilots.

In connection with this solution Professor Pugsley says:[5]

"The above process, though leading to a new logic of risk, which, as a philosophy, may supersede the comparatively irrational factor-of-safety system, has so far proved practicable only in relation to military aeronautics; and even there the data available,

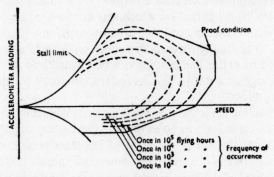

Fig. 163.—Frequency curves from V–g records.

[1] These stresses may amount to about 50 to 80 per cent.
[2] "Concepts of Safety in Structural Engineering", *J. Inst. C.E.*, London, March 1951, pp. 5 *et seq*.
[3] "The Safety of Structures", *Trans. Am. Soc. C.E.*, Vol. 112, 1947, p. 125.
[4] *Loc. cit.*, p. 18.
[5] *Ibid.*

though extensive, were really only sufficient to discuss accident rates in a relative rather than an absolute way."

Professor Freudenthal is more optimistic. In his concluding remarks[1] he expresses hope that statistical interpretation of involved influences will allow the elimination of arbitrarily specified limits, and thus achieve economy and, moreover, determine correct safety factors for unconventional structures embodying novel forms and materials.

Whether these conclusions are accepted or not, the method deserves attention as containing several original ideas. For instance, instead of referring to the factor of safety as a "factor of ignorance" (as was done earlier by various writers, see p. 193) Professor Freudenthal distinguishes between *ignorance* and *uncertainty*. The former term is meant to refer to our lack of precise knowledge about the laws of Nature controlling stresses, deformations, and other effects of mechanical loads, whereas the latter concept relates chiefly to inaccuracies of production and similar departures from the ideal, which are capable of being reduced by introducing quality control, standard tests, etc., but can never be removed entirely. A quantitative estimate of the effect of such inaccuracies on the required magnitude of the safety margin is then assumed to be capable of being arrived at by the methods of the mathematical theory of probability, as applied to past experience with similar works. For instance, in Fig. 164 Freudenthal gives, as an illustration to his ideas, the published results of acceptance-tests of structural shapes of silicon-steel for the towers of the Golden Gate bridge, in California.[2] The theoretical frequency curves fitted to the recorded deviations differ from the Gaussian curves in that they are skew curves, i.e. their "modes", or peaks, do not coincide with the mean values of the observed quantities. For specific conclusions, as derived from this figure, pertinent to the codification of tolerances, the reader is referred to the original publication.

It will be realized that Fig. 164 exemplifies the concept of "uncertainty" in relation to *production*. On the other hand, in Fig. 165 we have an instance illustrating the limitations of our basic knowledge (ignorance ?) concerning the *physical laws* of Nature. This is a composite diagram presenting three stages, or types, of interrelationships between slenderness ratios and buckling strengths of compressed members, which are selected for quotation as displaying "the simultaneous existence of functional, empirical, and statistical laws in one and the same problem".

Euler's ideal hyperbola based on theoretical elasticity appears in the *c* section of this diagram. On the other hand, in the *b* region we have the empirical law of inelastic buckling, whilst in *a*, failures occur altogether independently of the slenderness ratio. The point that the diagram is intended to make, is that in every one of these sections the grouping of the observation points is a statistical distribution, capable of being represented by a frequency curve—as shown in the drawing. Thus, although in the strict sense of the term the *b* stage alone is truly empirical, sections *a* and *c* must also be considered as a combination of functional and statistical relationships. In this sense, therefore, according to Freudenthal, "every physical property is a statistical distribution so that the constancy of physical qualities is found to be of a purely statistical nature".

It is fairly obvious that in the quantitative estimate of the safety margin, which is to be introduced in the design of structures, these various probable (or possible) departures from the respective ideal concepts must be superimposed upon one another; in other words, the final

---

[1] See footnote 3 on foregoing page.
[2] "Structural Applications of Steel and Light-Weight Alloys", a Symposium, *Trans. Am. Soc. C.E.*, Vol. 102, 1937, pp. 1328–29.

value of the factor of safety must reflect the amalgamated effect of all the individual "uncertainties" and "ignorances", on which it may eventually depend.

Here again we use the principle of mathematical expectation. In fact it will be remembered that according to the basic laws of the classical theory of statistics, the relatively slight

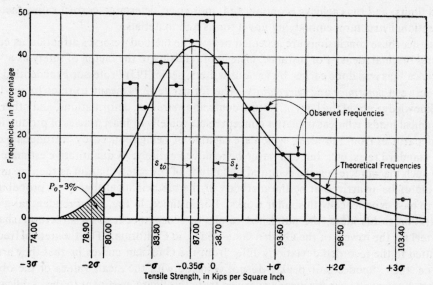

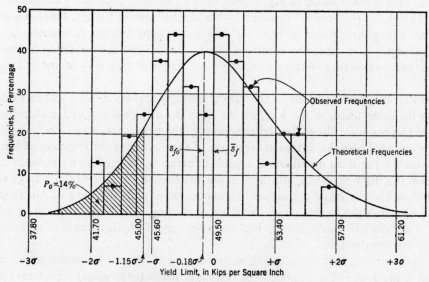

Fig. 164.—Comparison of observed and theoretical frequency distributions.

SPECIFICATIONS FOR STEEL BRIDGES

probability that all the component fluctuations will attain their maximum intensities at the same time, explains that the fluctuations of the result are materially narrower than might have been inferred from the summation of the components.

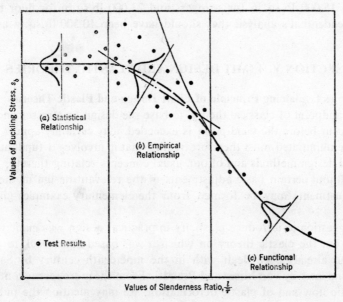

FIG. 165.—Buckling stress of steel columns.

Thus, according to the school-room theory of elementary probability, the fluctuations of the result are not the sum of the component fluctuations, but the square root of the sum of their squares, multiplied by the corresponding partial derivations.

Such a calculation is carried out separately for the "strains"—a comprehensive term used by Freudenthal to describe strains, stresses, failures, and all other mechanical results of application of loads—and for the influences affecting the "resistance", or "carrying capacity" or "strength", of the structure, and the maximum probable values of the former are then set against the minimum probable values of the latter, thus yielding a quantitative estimate of the probability of the occurrence of failure. In other words, the factor of safety is defined as the ratio of the minimum resistance to the maximum strain, and an outline is given of the methods to be used in carrying out the relevant calculations—the object being to exclude from this factor the personal, subjective element and base its estimate on purely objective consideration.

In examining critically this solution it is essential to realize that the terms *minimum resistance* and *maximum strain* as they appear in the foregoing paragraph, represent by assumption the average (or mode) values minus or plus the maximum range of fluctuation. This point deserves attention because the concept "maximum range of fluctuation" is not *entirely* scientific. It is *usually* taken as a multiple of the standard deviation—$j \times \sigma$ in Freudenthal's notation—in which the coefficient j is frequently assumed 3.5; but this is rather a *convention*—just what we aim to exclude.

On the whole, however, Professor Freudenthal's analysis of the effect of chance on the choice of the safety factor is pertinent information, for it is typical of the modern tendencies in structural philosophy.

From the purely practical point of view, he concludes to the advantage of a wider range of individual limits in the "graduated" or "impact" types of specifications. For instance, referring to the case illustrated in Fig. 164, i.e. the Golden Gate bridge, the "reduced" permissible limits were 15,000 lb/sq in for stringers and 23,000 lb/sq in for floor beams, whereas according to Freudenthal's analysis they should have been 10,500 lb/sq in and 27,000 lb/sq in, respectively.

SECTION V. LIMIT DESIGN AND PLASTIC THEORIES

1. Simple Examples Explaining Principle of Limit Design and Plastic Theories

From the standpoint of classical theory, to base the design on stresses and deformations, which will not occur before the yield-point is exceeded, may certainly appear an unorthodox proposition. The uninitiated must therefore realize that it involves a fundamental alteration of the traditional design methods and of our usual concepts relating thereto, and should not be attempted without certain basic adjustments of the relevant design methods. Some idea about these adjustments may be formed from the elementary examples given in this subsection.

The first suggestion to introduce plasticity in practical design was made by Kist, in Germany, in 1922,[1] but the plastic theory on which it was based dates back to a much earlier period, for it had already been dealt with in the nineteenth century by Saint-Venant and Maurice Levy, and later by von Mises and Prandtl. From the more recent contributions to the theories of plastic flow and of plastic deformation, we may mention the publications of A. Nádai,[2] W. Prager,[3] A.A. Ilyushin,[4] D.C. Drucker[5] and R. Hill.[6]

In England the application of the plastic theory to structural steel design is due chiefly to Professor J.F. Baker and his group at Cambridge University.[7]

The frequent references to the plastic theory and the so-called "Limit Design"[8] appearing in these pages may be made possibly more graphic and easier to visualize by the reader by means of a few simple examples,[9] which will explain these new concepts, in the light of the age-old philosophy of the "safety factor". Take, first, the simplest of all structural types—a uni-

---

[1] N.C. Kist, "Die Zähigkeit des Materials als Grundlage für Berechnung von Brüken, Hochbauten und ähnlichen Konstruktionen aus Flusseinen", *Der Eisenbau*, 1920, p. 425.

[2] A. Nádai, *Plasticity* (New York, 1931).

[3] W. Prager, "Ueber das Verhalten statisch unbestimmter Konstruktionen aus Stahl nach Ueberschreitung der Elastizitaetsgrenze", *Bauingenieur*, 1933, p. 65.

[4] A.A. Ilyushin, "The Theory of Elastic-Plastic Deformation and Its Application", *Bulletins of the Academy of Sciences* (Moscow, 1948), p. 769, and other publications by the same author.

[5] D.C. Drucker, "The Relation of Experiments to Mathematical Theories of Plasticity", *Journal of Applied Mechanics*, 1949, p. 349, and other publications.

[6] R. Hill, *The Mathematical Theory of Plasticity* (London, 1950).

[7] *Cf.* numerous publications by Professor J.F. Baker and J.W. Roderik; a list thereof will be found in the Bibliography annexed to Professor Baker's paper, "A Review of Recent Investigations into the Behaviour of Steel Frames in the Plastic Range", *J. Inst. C.E.*, London, Jan. 1949, p. 223.

[8] The term "Limit Design", in the sense here referred to, seems to have first been used by Professor J.A. Van den Broek in "Theory of Limit Design", *Trans. Am. Soc. C.E.*, Vol. 115, 1940, p. 638. See also: *Theory of Limit Design*, by the same author (John Wiley & Sons, New York, 1948), and numerous other publications on the subject.

[9] These examples, with drawings, etc., are abstracted from the paper by D.C. Drucker on "Plasticity of Metals—Mathematical Theory and Structural Applications", published in *Trans. Am. Soc. C.E.*, Vol. 116, 1951, pp. 1059 *et seq.*

SPECIFICATIONS FOR STEEL BRIDGES

formly loaded beam with fixed ends (see Fig. 166 (a)). So long as all the stresses are below the yield-point of the steel, the bending moments at the bearings, in A and B, will be twice as great as at the centre of the span, in C.

Supposing the load to go on steadily increasing, the limit of elasticity will first be reached (and exceeded) at the points where the bending moments are a maximum, i.e. at each end of the beam, in A and B. Assuming ideal plasticity to control further deformation in the overstressed sections, the moments of resistance at these sections will remain constant, say M_c; consequently, further increase of the load will not affect these moments, but in the middle of the span the bending moment will continue rising, its instantaneous value for a load p_i being $p_i l^2/8 - M_c$. This second stage of the case history of the beam is illustrated in Fig. 166 (b).

The third stage will be reached when $p_i l^2/8 - M_c = M_c$, for at that moment (see Fig. 166 (c)) the elastic limit in the middle point of the beam, in C, will be reached, and any further increment of the load will convert this point into another "plastic hinge" characterized by a constant bending moment, M_c.

At that stage again, the stresses will be capable of being easily assessed as shown in Fig. 166 (d). In fact, let us consider one-half of the beam, for instance, the left-hand one. It is subject to the effect of one-half of the distributed loading, i.e. $pl/2$, and of two moments M_c applied at each end of it. The shear will be zero in the centre of the span and $pl/2$ at the left end. All the elements required for the calculation of the beam are therefore at hand.

Thus, the principle of "limit design" yields a "statically determinate" solution to a "hyperstatic" problem, but the more important point of this elementary analysis is that it supplies a graphic demonstration of the fact that even after the elastic limit is exceeded the beam is nevertheless still capable of performing its duty. The "factor of safety" is thus presented in an entirely new light. One may only speculate as to how many beams and bridges, quite unknown to the authors of their designs, work under conditions similar to those of this example.

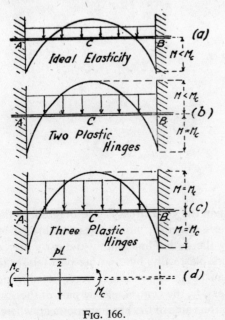

Fig. 166.

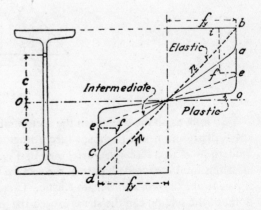

Fig. 167.

For instance, when the engineers of the Egyptian Bridges Service, in 1929, were instructed to calculate the stresses for the old Kazr-el-Nil bridge over the Nile, in Cairo, which dated back to 1870, they found that in the region of negative moments, i.e. over the supports, the stresses ranged from 3,000 to 4,000 kg/sq cm. That under such conditions this bridge was still capable of carrying the traffic, was due partly to the fact that instead of point forces, as assumed in the school-room theory of continuous girders, the reactions at the supports were distributed over a certain length, depending on the constructional features of the bearings; but was chiefly the result of the plastic yield in the critical spots, consequent on the particularly high ductility of the material, which, was of course iron and not steel.

Similar examples abound, but are not always as simple to detect, as may appear to be the case from the foregoing discussion, because usually, the respective effects of elasticity and plasticity cannot be as easily separated as we have assumed them to be, in investigating the beam shown in Fig. 166.

2. Deeper Inquiry into the Beam Failure

In fact, a deeper inquiry into the behaviour of the beam shown in this diagram would easily prove that our solution of the problem was oversimplified, by assuming that the deformation of the material in the sections A, B and C had *suddenly* changed from the elastic type, to the plastic regime. In Nature, this will not be so. The material in the outer fibres near the top and bottom of the beam will be overstressed first, and only then, the remaining, central part of the section will follow.

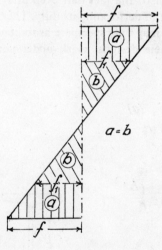

Fig. 168.

So long as all the stresses in the section are less than the yield stress f_y, the diagram of the stress distribution will be linear (according to Navier's principle, more or less confirmed by modern advanced research), as shown in Fig. 167 by the dotted line mn. In the early period after this limit is exceeded, part of the section will work plastically, but another part (closer to the neutral axis) will still remain elastic. This will then be an *intermediate* stage. Since in no point of the section can the stress exceed the yield stress, f_y, the central, elastic part of the section will be the only one to respond to the increase in the value of the bending moment, whereas in the extreme, plastic parts the stresses will be constant, and the general pattern of the stress diagram will therefore approximate two trapeziums, as shown in the same drawing by *bacd*.

As the external bending moment increases, the sizes of the plastic portions will grow until they very nearly meet in the centre of the beam and the entire section becomes plastic.[1] Only then will the moment of resistance become constant (as we have assumed it to be from the start) and equal to

$$M_c + 2A_1 c f_y$$

in which A_1 is the area of one-half of the section and c the distance between the centroid of this half and the axis (see figure).

The reader must, however, realize that the foregoing presentation of the case history of an overstressed beam gives only an outline of a rather involved phenomenon, some aspects of which are still subject to controversial opinion. For instance, there is the fact that the yield-point for steel in bending, is far higher than the yield-point for the same material when subject to axial tension.

3. W. Kuntze's Empirical Method and Other Theories

An empirical method devised by W. Kuntze[2] consists in considering the significant parameter for the critical conditions, at which the elastic limit is exceeded, not as the maximum stress f in the extreme fibres (see Fig. 168), but as the stress f_1 at the point which divides Navier's triangular stress diagram into two parts with equal areas. It can easily be shown that for a rectangular beam this solution furnishes a yield-point about 40 per cent in excess of that for uniform stresses, and this result agrees with the experiments of Thum and Wunderlich[3] and Eric Peterson.[4]

The difference between the conclusions based on uniform and non-uniform tension tests is particularly well shown in Fig. 169, which is abstracted from Mr. Eric Peterson's paper (p. 1201) and represents the results obtained with specimens made of the same material and tested in bending and in axial tension.

The lower curve in this diagram gives the strains obtained in the axial tension tests, while the deflections of the bending tests are shown by the upper curve. The load in bending at which the extreme fibres attain the yield-point in tension is 3,300 lb, as shown by 0. Had the yield-point in bending been the same, the deflection line above this point would have ceased to be straight, and must have followed the trajectory described as "old theory". But the tests showed that the line remained straight till the load reached 4,000 lb, giving a 25 per cent higher yield-point.

A most important conclusion, capable of being derived from M. Eric Peterson's results and other experiments on the same subject, is that mutual interaction between neighbouring

---

[1] The term "very nearly" as it appears in this sentence means that the stress at the neutral axis cannot change suddenly from positive to negative, because this would entail an infinite shear. There is therefore always a transition curve joining the positive and negative parts of the diagram, as shown in Fig. 167 by the curve *beed*.

[2] W. Kuntze, "Ermittlung des Einflusses ungleichförmiger Spannungen und Querschnitte auf die Streckgrenze", *Der Stahlbau*, Vol. 6, No. 7, 1933, p. 49.

[3] A. Thum and F. Wunderlich, "Die Fliessgrenze bei behinderter Formänderung", *Forschung auf dem Gebiete des Ingenieurwesen*, Vol. 3, 1932, p. 261.

[4] F.G. Eric Peterson, "Effect of Stress Distribution on Yield Points", Paper No. 2325, *Trans. Am. Soc. C.E.*, Vol. 112, 1947, pp. 1201 and foll. All relevant quotations as well as Figs. 168 and 169 are from this paper.

differently-stressed fibres has the effect of smoothing out, in the case of bending, the plastic irregularities of the stress-strain diagrams obtained from tests with uniformly stressed specimens. Better to grasp this effect, all the sharp corners in Fig. 167 were intentionally replaced by curves, which are intended to illustrate a principle, rather than certain specific test-results.

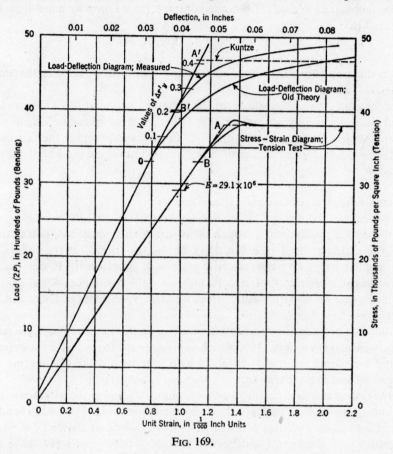

Fig. 169.

Still another assumption is illustrated in Fig. 167 by the line *ife*. It gives a general idea of the method of Professor J.F. Baker,[1] which is derived from a somewhat idealized relationship between stress and strain. The stress is assumed to rise to the upper yield-point of the material in *e*. When this point is reached, there is an immediate drop of stress to *f*.

Various methods have been suggested for estimating the moment of resistance during the transition stages, when part of the section is plastic, but another part is still elastic. To solve this type of problems as well as those of non-linear elasticity, A. Hrennikoff and others[2] use a method embodying the integral

$$m_1 = \frac{1}{\varepsilon^2} \int_0^\varepsilon \varepsilon f d\varepsilon,$$

[1] See footnote 7, p. 198.
[2] *Loc. cit.*, p. 169.

SPECIFICATIONS FOR STEEL BRIDGES

in which f and ε are the corresponding co-ordinates of the stress-strain function—whatever the latter might eventually be. The integral is then either calculated analytically, if this function is an equation, or by arithmetical summation, if the function is a composite empirical diagram. The value thus obtained represents the effect of the web, whereas the flanges are assumed to be subject to a uniform stress and appear, therefore, in the calculation as an area multiplied by a constant stress. Adding both parts together, we obtain for the section, inclusive of web and flanges, the formula

$$m = m_1 + kf = \frac{1}{\varepsilon^2}\int_0^\varepsilon \varepsilon f d\varepsilon + kf,$$

in which k is the ratio of the area of the flange to A_1. Thus, $k = A_f/A_1$ in which A_f is the area of the flange and A_1 is, as before, one-half of the area of the web.

Finally, the moment of resistance is found to be

$$M = 2mA_1 \times \frac{h}{2} = mA_1h.$$

The foregoing information is given here to demonstrate that, although the "limit design" approach to the problem of the safety factor broadens the outlook on the subject, it would be unwise to underestimate the mathematical difficulties associated with this solution. In fact, even in the simplest case of the beam we have considered earlier, in Fig. 166, it is not certain whether the sections A and B will have already attained the state of "ideal" plasticity before the section C is overstressed, and becomes therefore partly plastic. The ensuing conditions are obviously much more involved than what was assumed to have been the case in our first attempt at this problem.

4. Vertical Bar Flanked by Two Sloping Laterals

Another simple example which frequently re-appears in modern publications on the subject, is shown in Fig. 170 (see footnote 9, p. 198). Here we have a vertical bar, flanked by two sloping laterals and subject to a vertical load P.

The structure is a simple case of a statically indeterminate system, but as in the problem of the beam shown in Fig. 166, the limit-design-principle yields forthwith a statically determinate

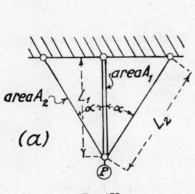

Fig. 170.

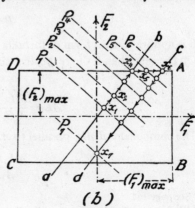

Fig. 171.

solution to this problem, for if we assume that the yield limit has been exceeded in all the three members of the system, we obtain, at once, the equation

$$P_{max} = f_y A_1 + 2 f_y A_2 \cos \alpha. \qquad (a)$$

Apart from the simple solution it yields, this equation deserves attention, also, because it may eventually lead to a more rational design, if contrasted with the usual working-stress limits, based on the classical, traditional interpretation of the safety-factor concept.

Viewed, however, from a more general standpoint, this novel interpretation of the safety factor is subject to certain criticism, for equation (a) as it stands, relates to stresses alone, but does not cover the problem of deformations, which, as we have already pointed out in connection with pre-stressed beams,[1] must not exceed a certain maximum, commonly fixed in proportion to the span.

We are, therefore, faced now with another problem, that of the plastic, or semi-plastic, deformation, which is not altogether obvious and is therefore another obstacle in the way of the application of the new theory. Attention is called in this connection to the fact that Hrennikoff's method, mentioned on page 169, contains a full solution of the problem of deformations, with tables, etc.

Reverting to Fig. 170, it may be used to explain the "shake-down" concept—another aspect of the same general theory, explored by W. Prager and P. Symonds.[2] Imagine a railway bridge subject to varying stresses, each panel load varying independently between a low and a high value any number of times, and suppose that over-stress occurred and plastic deformation took place in one or several points. It is required to find out whether, and when, the structure will adjust itself, i.e. "shake-down", so that it will never yield again under the same system of loads.

Better to grasp the significance of this problem, it may be useful to trace all the phases of the loading history of Fig. 170. So long as the load P is below the critical value which causes permanent plastic deformation, the stresses will be capable of being determined by means of the theory of simple elasticity, i.e. by solving two equations incorporating, respectively, the principle of the equilibrium of forces, and the principle of equal deformations.

To satisfy the requirement of equilibrium, the sum of the vertical components of the internal forces F_1 and F_2, acting respectively in the central bar and in the laterals, must be equal to the external load P. Hence, the equation

$$P = F_1 + 2 F_2 \cos \alpha \qquad (a_1)$$

which is a more general form of formula (a).

On the other hand, the respective elastic deformations of the main bar and of the laterals are

$$\Delta L_1 = \frac{F_1 L_1}{A_1 E_1} \quad \text{and} \quad \Delta L_2 = \frac{F_2 L_2}{A_2 E_2}.$$

The component of ΔL_1 parallel to ΔL_2, is $\Delta L_1 \cos \alpha$; hence the formula

$$\frac{F_1 L_1}{A_1 E_1} = \frac{F_2 L_2 / A_2 E_2}{\cos \alpha}.$$

---

[1] See page 164.

[2] W. Prager, "Problem Types in the Theory of Perfectly Plastic Materials", *J. Aeronaut. Soc.*, Vol. 15, 1948, p. 337.

Solving
$$F_2 = \left(\frac{A_2 E_2 L_1}{A_1 E_1 L_2} \cos \alpha\right) F_1. \qquad (b)$$

The graphical method is found to be most convenient for analysing these equations. Imagine, therefore, a system of orthogonal co-ordinates, with F_1 as abscissae and F_2 as ordinates, as shown in Fig. 171.

According to the common assumptions of ideal elasticity the maximum stress in the material must not exceed the plastic-yield stress f_y. The maximum values of the forces in the bars are, consequently, equal to
$$F_1 = \pm A_1 f_y = \pm (F_1)_{max} \quad \text{and} \quad F_2 = \pm A_2 f_y = \pm (F_2)_{max}.$$
These maximum values delimit the field investigated, and are represented in the diagram by the rectangle $ABCD$, within the boundaries of which the point representing the solution of the problem must of necessity be confined.

The deformation equation (b) is represented in such a diagram by the sloping straight line ab, whereas, for various values of the external, gradually increasing load P_1, P_2, P_3, etc., the equilibrium equation (a_1) yields a number of parallel lines with a negative slope $1/2 \cos \alpha$. The co-ordinates of the points of intersection of these lines with the deformation line ab, yield the forces F_1 and F_2, corresponding to different values of P; viz. as P increases the point of intersection rises, sliding along the line ab until it strikes the boundary DA in x_4. At this moment, the maximum limit of the elastic range is reached. If P increases beyond this limit (P_5 in the drawing) overstress takes place, the material in the lateral bars yields, and plastic deformation occurs. Hence, instead of continuing to move higher, following the same line ab, the intersection point begins to move horizontally, sliding along the limit DA (x_5 in the diagram).

For the sake of demonstration, let us assume that P_6 is the upper extreme limit of the range of variation of the load P, and that after reaching this maximum value, the load begins to decrease. Owing to the permanent plastic deformation, which has taken place whilst the point was moving from x_4 to x_6, the new deformation line will not be ab, but cd.

This latter line will, consequently, be the locum of the points representing the solution of the problem for the period in which the external load gradually reduces, and is finally reversed.

The described computation yields an obvious solution to the *shake-down* problem, for it is clear that the system will *shake-down* for values P_6 and P_7.

5. Rigid Horizontal Beam Suspended on Three Wires

It remains to explain another characteristic point bearing on the "limit-design" and plasto-elastic theory, viz. that in flagrant contradistinction to the rules of pure elasticity, strains and stresses are not more proportional to the loads, after the yield points have been locally exceeded and some of the constituent members of the system became, therefore, plastic.

Fig. 172 has been used[1] to demonstrate the point of this argument. Here again we have an elementary mechanical model consisting of a rigid member suspended on three wires of different lengths, areas and materials.

Within the limits of elastic deformation, the distribution of the load among the three wires in question conforms with the laws of elastic equilibrium. Hence

---

[1] From Mr. Drucker's paper, see footnote, p. 198.

$$\frac{F_1 L_1}{A_1 E_1} + \frac{F_3 L_3}{A_3 E_3} = \frac{2 F_2 L_2}{A_2 E_2}. \qquad (a)$$

In addition to which we shall also have two equations of static equilibrium, viz.

$$F_1 + F_2 + F_3 = P \qquad (b)$$

and

$$3 F_1 + F_2 - F_3 = 0. \qquad (c)$$

Note that the last two equations are valid under all circumstances and stress-strain interdependences.

We shall now suppose as in the earlier examples, that the load P gradually increases, and assume for the sake of argument, that the inner, longer wire is fabricated from a material (for instance, rubber) which is strong enough, but very flexible. On the other hand, the left-hand wire will be assumed to be much thinner than the right-hand wire.

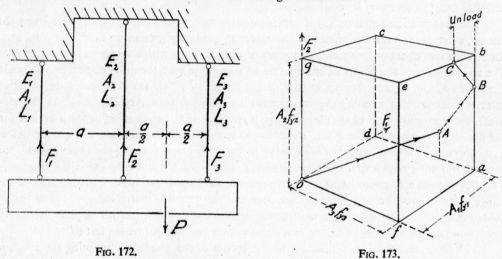

FIG. 172.   FIG. 173.

Solving, then, the three equations, we obtain

$$F_1 = \frac{P\left(\dfrac{2 L_2}{A_2 E_2} - \dfrac{L_3}{A_3 F_3}\right)}{2\left(\dfrac{L_1}{A_1 E_1} + \dfrac{4 L_2}{A_2 E_2} + \dfrac{L_3}{A_3 E_3}\right)} \qquad (d)$$

from which it is seen that F_1 is positive (tension) so long as

$$\frac{2 L_2}{A_2 E_2} > \frac{L_3}{A_3 E_3}.$$

For sufficiently small values of A_1 the stress in the left-hand wire will exceed the yield stress f_{y1} first. Further deformation of this wire will therefore conform with the principle of plasticity, and the maximum force it will be capable of carrying will be limited to $A_1 f_{y1}$.

Better to understand the load history of the three wires shown in Fig. 172 it will be convenient to adopt a graphical representation, following the same principle as in the foregoing example, with that difference, however, that instead of two variable quantities, we shall have to

SPECIFICATIONS FOR STEEL BRIDGES

deal, now, with three variables. Hence, instead of a two-dimensional diagram, we must here imagine a three-dimensional system of co-ordinates, as shown in Fig. 173.

The three planes *abcd*, *abef* and *bcge* delimit the elastic (or nearly elastic) domain, within which the point representing the solution, i.e. the three consistent co-ordinates F_1, F_2 and F_3 corresponding to a certain value of P, can move. Such movements are linear because the elastic deformation is proportional to the external load. Thus, as the load increases from zero to P_1, the point moves outwards along OA, which is a straight line. As the point reaches the limit A, in the diagram the force F_1 attains its maximum value $(F_1)_{max} = A_1 \times f_{y1}$, beyond which it cannot rise. This equation replaces, therefore, the formula (*a*).

As P increases beyond P_1, both F_2 and F_3 rise, but F_1 remains constant and equal to $(F_1)_{max}$. Hence, further movements of the point occur in the plane *abcd*, for instance as shown by AB. In drawing this line we have assumed that the next one to yield, is the third wire (point B). Thereafter the inner force in this wire will be $(F_3)_{max} = A_2 \times f_{y2}$.

Direct calculation will show that if P increases still further, F_1 must decrease, i.e. the left-hand wire will gradually be unloaded, as shown by the line BC.

Two wires have now yielded but failure does not occur because the central wire is still elastic, and the rigid bar rotates therefore clockwise, and thus causes the left-hand wire to unload.

In referring to this example, D.C. Drucker says[1]:

"The preceding discussion applies equally well to ideal plasticity or to the usual work-hardening stress-strain curves. Depending on geometry, and the type of these curves, the system will perform various tricks. For example, if the wires are really work-hardening bars capable of taking compression, the left-hand one may be made to yield in tension, unload, load in compression, yield, and unload again as the load P increases steadily."

Attention is called to two particularly important conclusions capable of being derived from this discussion. Firstly, that in elastic-plastic systems, deformations and displacements are not proportional to the loading, and secondly, that they depend on the consecutive order in which the loads are applied, i.e. on the "history" of the loading. In this, they differ essentially from the usual type of elastic systems familiar to the designing engineer.

SECTION VI. LIMIT STRESS FOR COMPRESSION

1. Danger of Buckling. Schwarz-Rankine Formula

It was explained in subsection 2, section I of this chapter, (pp. 166–167) that the problem of the choice of limit stresses to be inserted in the clauses of a specification for structural steelwork, i.e. bridges, gates, etc., was capable of being divided into two parts relating respectively to (*a*) the limits for tensile stresses, as the fundamental basis of the relevant theory, and (*b*) the ratio which other types of stress-limits bore to this fundamental basis. In the foregoing subsections we have dealt with problem (*a*). The second problem (*b*) will be considered in this and in the following subsections, beginning with the limit for compression.

Owing to the danger of buckling, compression members are calculated generally for a lower stress limit than that specified for tension members. The coefficient of reduction depends on the length versus the rigidity of the member, the latter being usually determined as a function of the least radius of gyration of the cross-section.

---

[1] *Loc. cit.*, p. 1071.

A particularly clear graphic idea of the behaviour of compression members under load is given by Professor Freudenthal in Fig. 165 *ante*.

The coefficient of reduction for structural steel-work is (or was) frequently calculated on the European continent according to the Schwarz-Rankine empirical formula, namely

$$\phi = \frac{1}{1 + 0.0008\frac{l^2}{r^2}}$$

where l is the length of the member and r is the radius of gyration $\sqrt{I/A}$.

The ratio l/r being a pure number, the formula will be the same, whether metric or English units are used.

2. Net and Gross Areas

In the earlier designs all the working stresses were usually calculated from the gross sectional area of the member. Subsequently, the net area became included in the calculation of the tension members, whilst compression members were still calculated according to the gross area. The present-day practice (at least on the Continent) is to calculate both compression and tension members according to the net area of the section. On the other hand, according to Tetmajer's experiments, the resistance of a riveted member against buckling may be calculated according to the gross moment of inertia, if the area of the rivet holes is less than 12 to 15 per cent of the gross area. Consequently, the calculation of a riveted member subject to compression will be made as follows: the working stress will be calculated from the net area of the section, whereas the coefficient of reduction ϕ will be calculated according to the gross area and gross moment of inertia.

3. Straight-line Formula

In America, the permissible compressive stress is often calculated according to "straight-line" formulae,[1] such, for instance, as the following:

$$f_{max} = 16{,}000 - 70 l/r \text{ lb/in}^2 = 1{,}125 - 49.2 l/r \text{ kg/cm}^2$$

with a maximum of $14{,}000 \text{ lb/in}^2 = 984 \text{ kg/cm}^2$.

The permissible tension stress had been assumed in this case to be 16,000 lb/in², a possibly too conservative, but very safe figure.

The coefficient of reduction was therefore

$$\phi = 1 - 70/16{,}000 \times l/r = 1 - 0.00437 l/r.$$

In applying such formulae the calculation of the compressive working stresses is based on the gross area and second moment of the section. The permissible limit, thus calculated, covers the sum of static and impact stresses.

4. Euler's Formula

The above-quoted buckling formulae are purely empirical. They are intended to be applied only when $l/r < 105$. For higher values of l/r, experiments show that the critical buckling

---

[1] It seems that the first straight-line formula was proposed by Thos. H. Johnson, in a paper to the Am. Soc. C.E., in 1886.

SPECIFICATIONS FOR STEEL BRIDGES

stress agrees fairly well with Euler's theoretical equation based on the principle of elasticity, namely

$$T = \frac{\pi^2 EI}{l^2}.$$

Accordingly, for such members the permissible compressive stress is:

$$f_{max} = \frac{\pi^2 EI}{NAl^2},$$

where A is the area and N the coefficient of safety.

For instance, in the old specification of the Prussian railways this coefficient was assumed to be 5.

More recent American substitutes for these various limits are as follows:

Axial Tension   18,000 lb per sq in
Compression:

(a) for $l/r < 140$ $15,000 - \frac{1}{4}\frac{l^2}{r^2}$ lb per sq in for riveted connections,

and $15,000 - \frac{1}{3}\frac{l^2}{r^2}$ lb per sq in for pin-connected members

(b) for $l/r > 140$ $18,000 - 5\frac{l^2}{b^2}$ lb per sq in,

in which $b = $ flange width.

Further information on the limits of application of Euler's formula, and on its relative value as compared with the various empirical "buckling" formulae, may be found in many text-books on Strength of Materials; but it is of limited significance only, insofar as practical bridge design goes. The reason is that the dynamic action of movable loads can produce buckling at a comparatively low stress intensity, owing to the individual vibrations of long and thin bridge members. This explains that in some specifications the maximum value of l/r for a compressed member carrying a calculated stress is limited to 100; consequently, in practice, the Euler formula will seldom be applied for the calculation of the main members of a bridge.

5. Failure of Old Quebec Bridge

The question of buckling stresses became a much-discussed topic, in the period following the failure of the Quebec bridge, in Canada, in 1907. This was an important bridge of the cantilever type, with anchor arms at either end and a suspended span in the middle. The size of the structure will be seen from the following figures: width of bridge 67 feet; central span 1,800 feet and two lateral spans 500 feet each. The bridge fell into the river, when one cantilever arm was nearing completion, killing eighty workmen. The cause of the failure was traced to insufficient stiffness of the bottom compression chords, near the piers. Prompted by public interest, which centred on this failure, several investigations on the buckling of structural members were then published by different authors, and much new light was thus thrown on the subject. From among other works, Ostenfeld's, Karman's and Krohn's investigations based on Tetmajer's experiments, deserve to be particularly mentioned. The conclusion arrived at was that the resistance against buckling of a member consisting of several individual parts braced together, was less than might have been expected, judging by the combined moment of inertia of the constituent parts, referred to the same neutral axis.

6. Buckling of Upper Flanges in Plate Girders

It might be of interest to observe that, with reference to compressive stresses in plate girders, continental European design-practice differs essentially from the methods adopted in America. In Europe the same stress intensity is allowed for the upper and lower flanges, whereas in America, the compression flange of a plate girder is designed to resist buckling, on the same basis as a compression member of a lattice girder. For instance, in one of the specifications of the A.R.E.A. the permissible unit stress for the compression flange was $16,000 - 200\, l/b$ lb/sq in, when the flange consisted of angles and plates, and $16,000 - 150\, l/b$ lb/sq in when it was built up of channel sections. In this formula l and b denote respectively the horizontally unsupported length and the width of the flange. The stress is supposed to be calculated according to the gross area of the flange.

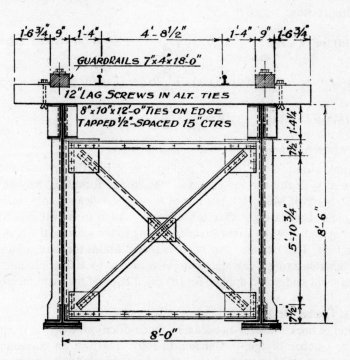

FIG. 174.—Section of a typical unsymmetrical American bridge.

On the other hand, on the European continent the plate girder is usually assimilated to a rolled joist, and the permissible limit for compression is, consequently, assumed equal to the limit for tension, due allowance being made, in either case, for rivet holes.

This explains the fact that the standard plate girder on the Continent is symmetrical, whereas in America the upper flange differs substantially from the lower flange (see, for instance, Fig. 174). Further information on this point will be given in the pages dealing specifically with plate-girder designs, as such.

SPECIFICATIONS FOR STEEL BRIDGES

SECTION VII. LIMITS FOR SHEAR AND BEARING STRESSES

1. Shear

Let f_t be the permissible tensile stress and let f_s be the permissible shearing stress. The ratio f_s to f_t, as adopted in different Continental specifications, is (or was) as follows:

Austria	Railway bridges	$f_s = 0.70 f_t$
France		$f_s = 0.75 f_t$
Russia	Railway bridges:	
	for the web	$f_s = 0.75 f_t$
	for rivets	$f_s = 0.80 f_t$
Switzerland		$f_s = 0.90 f_t$
Germany	Railway bridges	$f_s = 0.90 f_t$

In America, different permissible shearing stresses are allowed for shop- and field-driven rivets, respectively. For instance, according to one of the A.R.E.A. specifications the permissible limits for shear were as follows:

For shop-driven rivets	$f_s = 0.75 f_t$
For field-driven rivets	$f_s = 0.625 f_t$

At one time, much attention was devoted to the additional strength of riveted connections afforded by the friction force developed between the contacting surfaces of the different members in a riveted joint. At present, however, the usual practice is to neglect this friction entirely, owing to its uncertain nature.

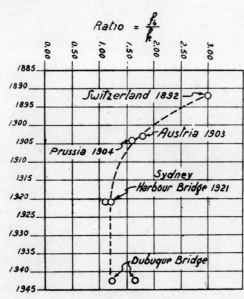

FIG. 175.

It must be borne in mind that when the graduated type of specification is used, the shear limit for rivets connecting two or more members of different classes of steel, must be determined with reference to the lowest tensile limit permissible for these various classes.

It is a common opinion among designers that riveted connections are, possibly, the less resistant part of all the steel structure. Consequently, there seems to be a more or less openly avowed general tendency to increase the safety factor incorporated in such connections. For instance, in the case (which often occurs in bridge design) where a member possesses a greater resistance than the calculated stress, the riveted connection is frequently specified to be designed according to this greater resistance; indeed, such a clause forms part of many modern specifications.

2. Bearing Stress

From among the various stress limits controlling the design of modern steel-work, the bearing-stress limit is the youngest. It first originated in America, but migrated later to England and Germany. The Latin countries of Europe were the latest to recognize its practical significance.

Diagram Fig. 175 shows characteristic instances illustrating the historical growth of the importance of the bearing-stress limit as a practical criterion for design.

It will be observed in inspecting this diagram that, in the earlier period, the relative value of the bearing limit was so high that it could scarcely affect the design of calculated sections. Subsequently, however, the criterion became gradually more severe (and therefore significant), and since about the year 1920, an almost constant ratio between the bearing and the tensile limits seems to have been permanently established.

Chapter Seven
PLATE-GIRDER BRIDGES

SECTION I. GENERAL ORDER OF PROCEDURE

In designing a bridge we must necessarily begin with the members directly supporting the live loads, and finish with the stresses on the foundation; that is to say, we follow the same consecutive order in which the stress due to the live load is transmitted from member to member of the structure.

Nevertheless, when calculating the strength of the different parts of the anatomy of a bridge, it is first necessary to work out a preliminary sketch of the entire work, before we can start the final calculation; for instance, before calculating the strength of the rail bearers, it is first necessary to figure out their theoretical span, which is equal to the length of a panel. As all the panels must of necessity be equal, their length depends on the exact distance between the centres of the bearings at either end of the bridge, which distance is usually referred to as the "theoretical" span. On the other hand, if the clear opening is given (as is usual), the theoretical span depends *inter alia* on the size of the bearing plates, and the latter are calculated according to the total reaction, inclusive of live load and over-all weight of the span. It will thus be seen that finally, the span of the rail bearers depends on the total weight of the bridge. Accordingly, before we can begin the calculation of the rail bearer, it is first required to estimate approximately the total weight of the bridge, and to fix its general layout. This is done by comparison with existing bridges of more or less the same type. To facilitate this calculation, a variety of empirical formulae have been suggested, giving the ratio of the theoretical span to the clear span, as well as the structural weight of the bridge. As two designs are never exactly the same, it is advisable to verify the results obtained by application of one formula, by means of another equation.

The theoretical span may be estimated either directly, or through the calculation of the required areas of bearings, according to the approximate weight of the bridge and the live reaction.

The following empirical formula, giving directly the theoretical span as a function of the clear span, may be used:

$$l_t = 1.01 l_c + 0.45 \text{ metre},$$

when l_c is below 20 metres and

$$l_t = 1.017 l_c + 5/l_c,$$

when l_c is above 20 metres.

In this formula l_t and l_c are the theoretical and the clear spans of the bridge (in metres).

For the probable weight of plate-girder railway bridges Dirksen gives the following formulae.[1]

Type I. Open-floor Deck Bridges. The weight of the main girders, wind bracings, bearings, etc.: $240+54l$ kg per linear metre of bridge. Weight of sleepers, rails, etc., from 640 to 775 kg per linear metre of bridge. Total weight from $880+54l$ to $1,015+54l$ kg per metre run of bridge.

Type II. Open-floor Trough Bridges. Weight of main girders and wind bracings, $270+44l$ kg per linear metre of bridge.

The weight of rail bearers, cross-girders, etc., is given in proportion to the spacing of main girders, as follows:

Spacing of main girders, in metres	3.0	3.3	3.7
Weight, in kg per metre run of bridge	380	430	520

The weight of rails, sleepers, bolts, etc., is 595, 630 or 660 kg per metre run, depending on the spacing of the main girders, as above. Accordingly, the total weight of bridge is $1,245+44l$, $1,330+44l$ and $1,450+44l$ kg per linear metre, when the spacing of the main girders is 3.0, 3.3 and 3.7 metres respectively.

For intermediate spacings of girders, the respective constants are to be found by interpolation. On railways where the weight of the rails, bolts, sleepers, etc., is standardized, the corresponding constants must be adjusted accordingly.

Type III. Ballasted Flooring, Trough Bridge. Weight of main girders including wind bracings, per metre run of bridge: $270+49l$ kg.

Weight of cross-girders and joists 670 or 840 kg per metre run of bridge, if the spacing of main girders is 3.3 or 3.7 metres respectively; the weight of sleepers, rails, ballast, etc., is 2,840 to 3,260 kg per metre run of bridge, depending on the spacing of the main girders.

The total weight of *Type III* bridges per metre run of bridge is $3,780+49l$ to $4,370+49l$ kg depending on the same spacing.

The above-quoted formulae are based on German practice. The American standards may be illustrated by means of the following figures abstracted from the tables published by Milo S. Ketchum (Illinois Central Railroad):

Trough Plate-Girder Bridges (Type II)

Span in feet	30	35	40	45	50
Spacing of main girders	15' 6''	16' 6''	17' 6''	17' 6''	17' 6''
Total weight of span in lb	45,000	56,000	64,000	71,000	81,000

Deck Plate-Girder Bridges (Type I)

Span in feet	30	35	40	45	50
Spacing of main girders	7' 0''	7' 0''	7' 0''	7' 0''	7' 0''
Total weight of span in lb	18,000	22,000	28,000	34,000	40,000

Constants for intermediate span values are obtained by interpolation.

The preliminary estimate of the theoretical or effective span, by means of the above formulae and figures, is made as follows.

Let the total weight of the span be W. For simply supported beam bridges the dead-weight reaction per bearing is $W/4$. The approximate maximum live-load reaction, per girder, may be easily calculated, assuming the effective span to be equal or slightly greater than the clear span. Let this reaction be T. The total reaction is $T+W/4$. Let now the permissible bear-

---

[1] In the following formulae l is the span in metres.

PLATE-GIRDER BRIDGES

ing stress on masonry be f (from 5 to 10 kg per sq cm). The required area of the bearing stone will be:

$$A = \frac{T + W/4}{f}.$$

Assuming the bearing stone to be square, the side of the stone is

$$b = \sqrt{A} = \sqrt{\frac{T + W/4}{f}}.$$

The bearing stone is preferably placed at some distance a (see Fig. 176) from the face of the abutment so as to prevent the crushing of the edge of the masonry. The distance a for small bridges may be about 10 cm. For large bridges it may rise to as much as 50 cm.

With reference to the sketch in Fig. 176 the theoretical span is

$$l_t = l_c + 2a + \sqrt{\frac{T + W/4}{f}}.$$

The results of this formula must be adjusted to a round figure taking into account the number of panels, in such a manner as to avoid awkward figures for panel lengths.

For instance, the average panel length in the plate-girder bridges on the Egyptian State Railways ranges from 2.5 to 3.5 metres.

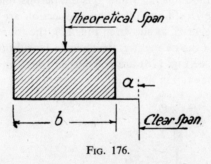

Fig. 176.

The axis of the last cross-girder must fall exactly in line with the centroids of the bearing plates.

From the information supplied above, the preliminary plan of the bridge may be worked out. In sketching the approximate cross-section and elevation of a railway bridge, the height of the plate girders may be taken as about one-ninth to one-eleventh of the span, and slightly less, for a highway bridge. If this criterion is not satisfied, special means must be taken in order to reduce the deflection, that is to say, extra metal is provided to augment the moment of inertia, so as to increase the rigidity of the bridge, even though that extra metal may not be necessary insofar as stresses are concerned.

SECTION II. PLATE-GIRDERS IN GENERAL

1. Two Types: Continental and Anglo-American

Before we can continue the discussion on railway bridges *in particular*, it will be advisable to give some explanatory notes on plate girders *in general*.

In the first instance, it is important to understand that two different methods may alternatively be followed in designing a plate girder. On the European Continent, the design is based on the moment of inertia of the cross-section, whereas in Britain and America the plate girders are more often designed by the flange area method.

2. Continental Method

Let I_{gr} be the gross moment of inertia ("second moment") of the normal section of the plate girder, and let I_r be the combined moment of inertia of the areas of the rivet holes. The figure to be included in the calculation is the net moment of inertia, namely

$$I_{net} = I_{gr} - I_r.$$

The bending stress which controls the design is obtained from the usual bending-stress formula

$$f = My/I_{net}$$

where M is the bending moment, at the critical sections of the girder, and y is the distance from the neutral axis to the extreme fibres of the material (see Fig. 177). The same permissible stress is allowed for tension and compression flanges; consequently, the moment of resistance of the girder is

$$N = Z_{net} \times f_t$$

where Z_{net} is the net section modulus I_{net}/y and f_t is the permissible tensile stress.

As a general rule, all the rivet holes falling in one section are to be subtracted in calculating I_{net}. If the rivets are staggered, as shown in Fig. 177, the four vertical rivets a—a must be included in the calculation of the moment I_r which is to be subtracted from I_{gr} to obtain I_{net}. The horizontal rivets b—b (see Fig. 178) are considered only for plate girders without cover plates, or in the case when they fall in line with the vertical rivets a—a.

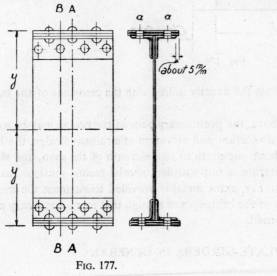

Fig. 177.

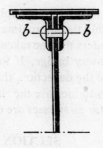

Fig. 178.

The section of the plate girder, designed according to this method, is usually symmetrical. The flange plates are about one centimetre wider than the sum of the horizontal legs of the

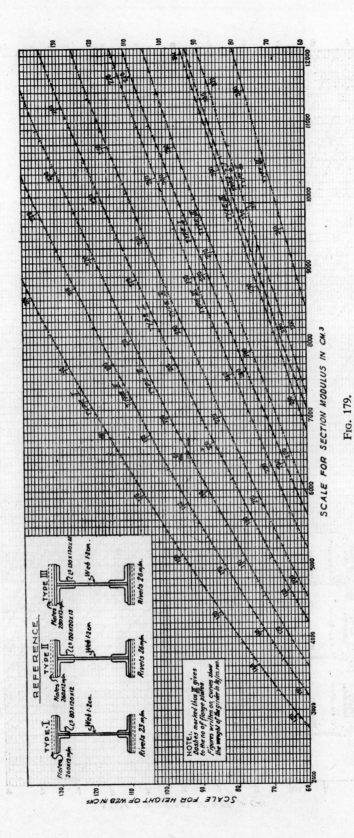

Fig. 179.

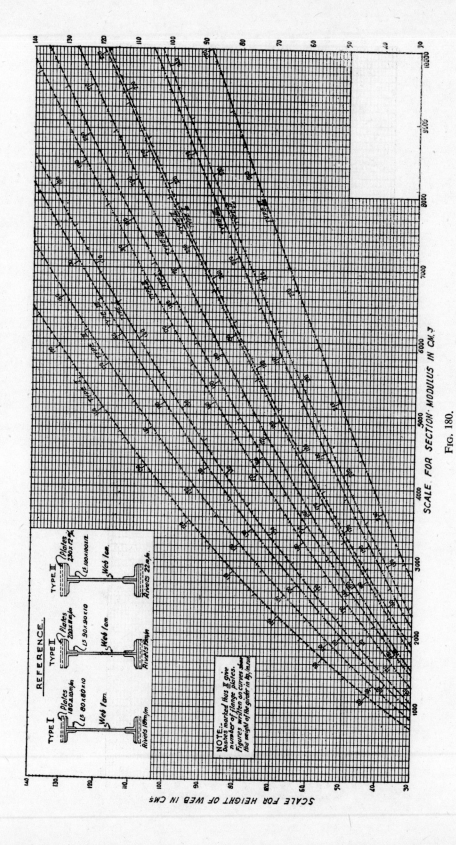

Fig. 180.

PLATE-GIRDER BRIDGES

flange angles plus the thickness of the web. A margin of about half a centimetre is thus provided, on either side of the flange, so that the edges of the angle-bars (which may not be perfectly straight) should not project beyond the edges of the flange plate. Wider plates must preferably be avoided, unless additional rows of rivets are arranged in the projecting portions. Otherwise, the plating shows a tendency to open up at the edge, thus giving access to moisture, rust, etc. Critics of this system of design allege that the calculation of the moments of inertia is a lengthy operation, resulting in a considerable loss of time; especially when the final result is obtained after several preliminary trials. It must be pointed out, however, that the use of the moment of inertia method is greatly simplified by tables, giving directly the net section moduli for different types and heights of plate girders. When such tables are available this method will be more expedient than the area method. The diagrams in Figs. 179 and 180 (thus far unpublished) permit the method to be applied with even greater ease, than the tables.

The curves in these diagrams give the section moduli of frequently used types of plate girders. The ordinates represent the height of the web, and the abscissae are the corresponding net section moduli. For sections without flange plates, the net modulus is calculated on the basis of subtracting the horizontal rivets connecting the flange angles to the web. For girders provided with cover plates, the vertical rivets are taken into account, assuming that the horizontal rivets do not fall within the same section. The figures on the curves give the weights of the girders in kilograms per metre run. In Fig. 179 the thickness of the web is 10 millimetres, and in Fig. 180, 12 millimetres. The scantlings of the sections are adjusted accordingly.

By means of these diagrams the design of the plate girder is greatly simplified.

Let ABC in Fig. 181 be the bending-moment diagram for the bridge, and let X be the maximum moment. In this drawing it will be seen that the maximum moment is given by the ordinate BD. The corresponding section modulus is $Z_{net} = X/f_t$, where f_t is the permissible stress for tension. Should there be no particular reason for doing otherwise, the height of the web, h, would be selected as from one-ninth to one-eleventh of the span. The values of Z_{net} and h will determine a point on either Fig. 179 or Fig. 180. Choosing one of the curves passing near this point, we obtain the size of angle-bars, flange plates and rivets. The height of the web h is then adjusted so as to give Z_{net} equal to or slightly greater than the value required by the calculation.

Suppose that the section, thus determined, includes two plates for each flange. As the moment near the supports is much smaller than at the centre, there is no need to extend these

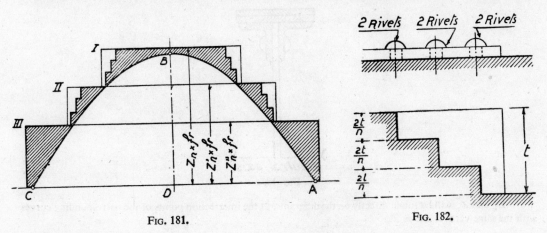

Fig. 181.   Fig. 182.

plates throughout all the length of the girders. The problem may then be solved graphically by plotting on the bending moment diagram in Fig. 181 the lines *I, II* and *III* corresponding to the values of $Z_n \times f_t$, $Z'_n \times f_t$ and $Z''_n \times f_t$ respectively, Z'_n and Z''_n being the section moduli for one flange plate per flange, and no flange plate, respectively.[1]

Referring again to Fig. 181, the *theoretical* lengths of the flange plates is given by the intersection of the horizontal lines *II* and *III* with the curve of the bending moment diagram.

Practically, the cover plates must be made slightly longer; namely as follows: let the sectional area of the plate be a and the diameter of the rivet d. The total stress which the cover plate may eventually carry is $a \times f_t$. The number of rivets required to transmit this stress is:

$$n = \frac{a \times f_t}{\pi d^2/4 \times f_s},$$

where f_s is the permissible limit for shear, in the rivet. Every pair of rivets corresponds to two nth parts of the total strength of the flange plate; consequently, the diagram of the moment of resistance of the end of the flange plate will be a stepped polygon (see Fig. 182), the rise of every step falling in line with the axes of each pair of rivets.

In Fig. 182 the number n is supposed to be 6. There are consequently three steps in this diagram. In this case, therefore, every pair of rivets engages one-third of the area of the plate. Thus, the full strength of the flange plate is developed to the left of the third row of these rivets.

It is evident that the flange plates must be of such lengths that the stepped diagram representing the moment of resistance (hatched in the sketch) should not intersect the line representing the bending moment.

If this criterion is satisfied, the actual bending stress M/Z will be, at any point of the girder, less than the permissible stress, f_t.

Applying the moment of inertia method, the shearing stress in the web is figured out according to the same formula, which is used for rolled joists, namely

$$f_s = QS/Ib,$$

where Q is the shear and S the first moment (about the neutral axis of the beam) of the sectional area located beyond the point for which the stress is calculated. The letter b here repre-

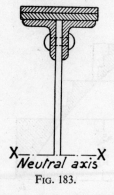

Fig. 183.

[1] Z_n and Z''_n will be found directly on the diagrams, at the intersection points of the corresponding curves with the same vertical line as Z_n.

PLATE-GIRDER BRIDGES

sents the thickness of the web. The proof of this formula will be found in every college course on the Strength of Materials.

The pitch of the rivets is determined from the following considerations.

The horizontal rivets (b—b in Fig. 178) resist the shearing stress tending to separate the flange from the web, namely, the shearing force acting on the area shown hatched in Fig. 183. According to the common theory of bending, this force, per unit length of the girder, is

$$T = QS_f/I,$$

where S_f is the first moment of the flange (shown hatched on the sketch) about the neutral axis XX. Let p be the pitch of the rivets. The number of rivets per unit length is $n = 1/p$. Hence, the stress acting on one rivet is

$$t = T/n = T_p = QS_f p/I. \qquad (A)$$

The horizontal rivets b—b are working in double shear, consequently, in terms of the shearing stress, the pitch of the rivets must satisfy the following criterion:

$$f_s \geqslant \frac{2QS_f p}{I\pi d^2} \quad \text{or} \quad p \leqslant \frac{f_s I \pi d^2}{2QS_f}.$$

On the other hand, considering as criterion the maximum permissible bearing stress, the pitch will be found from the formula:

$$f_b \geqslant \frac{QS_f p}{Idb} \quad \text{or} \quad p \leqslant \frac{f_b b I d}{QS_f},$$

where b is the thickness of the web.

The strength of the vertical rivets may be checked in the same manner, substituting for the first moment of the flange, the first moment of the flange plate, alone. Verification of the pitch of vertical rivets is required only in the case if it is greater than the pitch of the horizontal rivets. Otherwise the vertical rivets are stressed less than the horizontal rivets.

As an approximate empirical rule, the pitch of rivets in plate girders is about $4d$. If the calculation shows a pitch smaller than $3.5d$, it is preferable to adopt angle bars equal to or larger than 12.5 cm by 12.5 cm (5 × 5 inches); in which case we may arrange the rivets in two rows. The permissible limit for the pitch will then be doubled.

3. Anglo-American Method

The design is based, in this case, on certain conventional assumptions. The details of these assumptions vary from specification to specification, but the general lines of the method are as follows:

(1) The flanges alone are supposed to take all the bending effect, whereas the web is supposed to carry only the shear.

(2) The area of the flanges is supposed to be concentrated in its centroid, the stress being uniformly distributed over all the flange area. If h is the vertical distance ("effective depth") between the centroids of the upper and lower flanges and R represents the total force acting on one flange, then the moment of internal stresses is $h \times R$. This moment must be equal to the external bending moment M. Consequently $R = M/h$, and the normal unit stress on the flange area is

$$R/A = M/hA.$$

The shear is suppsoed to be taken by the web; consequently, the intensity of the average shearing stress is Q/a, where a is the sectional area of the web.

(3) The gross area is included in the calculation of the compression flange and the net area is used in calculating the tension flange. Apart from that, different permissible limits are frequently allowed for compressive and tensile stresses respectively. Thus, if A' is the area of the tension flange, and A'' is the area of the compression flange, the two following criteria must be satisfied at every section of the girder:

$$f_t \geq \frac{M}{hA'_{(net)}} \quad \text{and} \quad f_c \geq \frac{M}{hA''_{(gross)}}.$$

Hence, the bending moment diagrams must be plotted twice (as against one diagram only, in the first method). In harmony with this double design, the flanges are often made different, if the "flange area method" is applied.

A common type of flange arrangement, frequently used on English railways, is shown in Fig. 184 (a). It will be observed that the flange plates are here made much wider than in the continental type (see Fig. 177), and consequently, four rows of vertical rivets are usually placed in each flange. The ends of the stiffeners are bent and riveted to the flange plates; without this, the plates would have lacked in stiffness, owing to their great width.

The very wide flanges of the English type of plate girder are well suited to resist the buckling effect, but the bent stiffeners require specialized workmanship, and are therefore expensive. Apart from that, the smaller the area of every flange plate, the closer the approximation to the ideal shape of the curved beam of equal resistance.

On the other hand, the sketch b in the same drawing, is a typical American design.

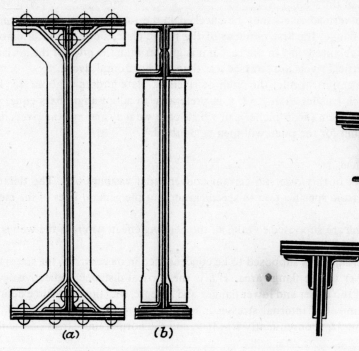

Fig. 184. Fig. 185.

PLATE-GIRDER BRIDGES

4. Comparison of the Two Types

Ceteris paribus, the maximum span of a plate-girder bridge is not a purely technical, but rather, an economical consideration, for it depends on the relative costs of labour and material. After a certain limiting span is exceeded, a region is reached in which both the plate girder and the lattice girder are possible from the structural standpoint; but the former will always involve more material and less labour, whereas the second will be lighter, but more expensive to construct and erect.

This explains why it is that in America, where labour cost is relatively high as compared with the cost of material, plate girders have for a long time been a popular type, even for large spans.

For such a bridge, a flange consisting of two angle-bars and three cover plates becomes insufficient. On the other hand more than three cover plates must not be used, because the vertical rivets would then be too long, while the horizontal rivets would carry an excessive shear. Accordingly, special types of flanges have been developed in America, with a much greater area than the type adopted in Europe. Instances of such flanges are shown in Fig. 185.

The lengths of the flange plates in such designs are determined separately for the top and bottom flanges. Two diagrams, similar (except for the stepped extremities) to the diagram in Fig. 181, are then computed, for the lower and upper flanges respectively (see Fig. 186).

The first diagram represents the bending moment superposed upon the values: $(A'_{net}) \times f_t \times h$. The second diagram gives the bending moment and the values: $(A''_{gross}) \times f_c \times h$. The extra lengths, l, of the flanges are supposed to contain enough rivets to engage the full resistance of each plate.

As already stated, the flange-area method varies, in its details, from one specification to another. For instance, it is assumed in several American specifications that one-eighth of the web is included in the calculation of the flange area. On the other hand, there are English designs in which only the angle-bars and the flange plates are considered as flange area, the same permissible stress being adopted for the top and bottom flanges. In the latter case, the rivet holes are deducted in the calculation of both flanges.

In comparing the two alternative methods of plate-girder design, the following points call for particular mention: in the "moment of inertia" method the bending stress is calculated according to the formula $f = My/I$, while in the "flange area" method the stress is found from the formula $f = M/hA$, which may be written as

$$f = M/hA = \frac{M\left(\dfrac{h}{2}\right)}{\left(\dfrac{h}{2}\right)^2 2A}.$$

Thus, the assumptions made in applying the second method are as follows:

I is assumed to be equal to $2A\,(h/2)^2$

y ,, ,, ,, ,, ,, ,, $h/2$.

Since these are reasonable approximations, it follows that the fundamental basis of both methods is the same. In other words, the difference lies in the application of the theory, but not in the theory itself, and the analogy can therefore be continued still further; for instance, including the approximate value of I in the formula for the shearing stress, and replacing, accordingly, S by $A \times h/2$, we get

$$f_s \geq QS/Ib = \frac{QAh/2}{2A\,(h/2)^2 b} = Q/hb,$$

which is, indeed, the formula applied in calculating the shearing stress as per the "flange area" method. As h approaches closely the geometrical depth of the web, we may write (approximately):

$$f_s \geq Q/a,$$

where a is the area of the cross-section of the web.

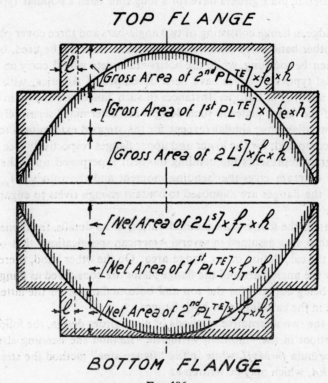

Fig. 186.

The formula for the pitch of the rivets may be found from similar considerations.
The stress (per unit length of beam) tending to separate the flange from the web will be

$$T = QS/I = \frac{Q\,(h/2)\,A}{2\,(h/2)^2\,A} = Q/h.$$

If p is the pitch of the rivets and l/p is the number of rivets per unit length of the girder, the stress carried by one rivet is

$$t = Tp = Qp/h.$$

Following the same path of reasoning,

$$f_s \geq \frac{Qp}{h^2 \pi d^2/4} \quad \text{or} \quad p \leq \frac{f_s \times h \times \pi \times d^2}{2Q}$$

PLATE-GIRDER BRIDGES

$$f_b \geqslant Qp/hbd \quad \text{or} \quad p \leqslant \frac{f_b \times h \times d \times b}{Q}.$$

These formulae may be interpreted as follows: the flanges of the plate girders must be fixed to the web by a sufficient number of rivets to transmit the total shear, at any point, over a distance equal to the effective depth of the girder at that point (see Clause 31 of the A.R.E.A. specification).

Further information on the design of plate girders will be given further on, in the section devoted to the details of plate-girder bridges.

SECTION III. RAIL BEARERS AND CROSS-GIRDERS

1. Design of Rail Bearers

The object of the rail bearers (stringers, joists, secondary longitudinal members; French *longerons sous voie*) is to support the sleepers and to transmit the load of the train to the cross-girders. The question of the span of rail bearers may be solved on the basis of minimum cost; namely, the greater the spacing of the cross-girders, the greater the weight of the rail bearers (per unit length of bridge), but the less the number of cross-girders. Consequently, a certain span of rail bearers gives a minimum structural weight. Broadly speaking, the economical span of the rail bearers is of the order of the spacing of the main girders, i.e. from 2.5 to 3.5 metres, for single-line railway bridges of normal gauge. It should be noted, however, that for lattice girders the proper selection of the span of rail bearers is a much more involved problem, depending on the type of the main trusses adopted.

The rail bearers may either be rolled joists or plate girders; but, if the latter type is adopted, the cover plate (if any) should be extended over the full length of the bearer, because all the sleepers must be placed at the same level. Consequently, the rail bearers are most often made of uniform section, and are designed according to the maximum bending moment. The position of the locomotive giving the maximum bending moment is found according to the general principles of the theory of structures.

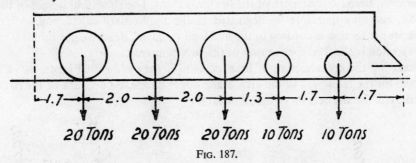

FIG. 187.

Example. Find the maximum bending moment for a rail bearer, the spacing of the cross-girders being 3.20 metres, and the standard locomotive being as shown in Fig. 187.

1st Case. Two large wheels on the rail bearer (see Fig. 188). To get the maximum bending moment the wheels are located in such a manner that the distances from the centre of the span to the largest wheel and to the centroid of the two wheels, are equal:

$$M_{max} = 1.1 \frac{10 \times 2.1 + 10 \times 0.1}{3.2} = 7.56 \text{ ton} \times \text{metres}.$$

2nd Case. One large wheel on the bearer (see Fig. 189). The wheel is placed in the centre of the span:

$$M_{max} = 1/4 \times 10 \times 3.2 = 8.00 \text{ ton} \times \text{metres}.$$

3rd Case. A large wheel and a small wheel on the bearer (see Fig. 190). The critical position is found according to the same principle as in the first case. The centroid of the two wheels is

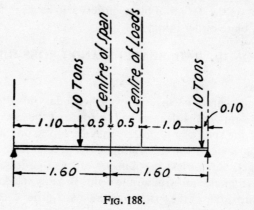

Fig. 188.

at $\dfrac{5 \times 1.3}{10+5} = 0.433$ metre from the large wheel; consequently, the large wheel is located at $0.433/2 = 0.217$ metre from the centre point of the span:

$$M_{max} = 1.383 \, \frac{10 \times 1.817 + 5 \times 0.517}{3.2} = 8.98 \text{ ton} \times \text{metres}.$$

Thus, the *maximum maximorum* moment is found in the third case.

In designing the section of a rail bearer, the impact coefficient is practically equal to one hundred per cent, because the length of the bearer is short. The structural weight of the bearer is very small, and may therefore be neglected in the preliminary calculation. After having selected the proper section according to the live-load moment, dead-weight moment is figured out and the section is checked for the sum of the two moments.

In calculating the cross-section of bearers, even if they are rolled joists, due allowance must be made for the reduction of the area of the upper flange, on account of the rivets which are required to fix the sleepers on to the rail bearers.

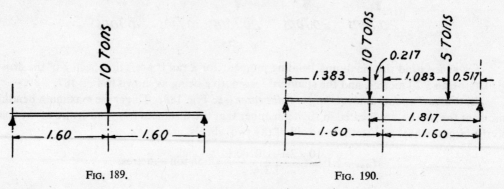

Fig. 189. Fig. 190.

PLATE-GIRDER BRIDGES

This may be done in several different ways. Fig. 191 shows an arrangement frequently used in Europe. A heavier device of the same type consists of two short angles, one on either side of the sleeper, with one single bolt joining the vertical legs of the two angles. If there is no

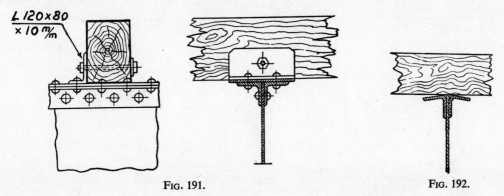

Fig. 191.   Fig. 192.

cover plate on the rail bearers, a small base-plate is placed under the sleeper. This is done to prevent the opening of the vertical joint under the sleeper as shown in Fig. 192.

Fig. 193 shows the arrangement adopted, for fixing the sleepers, on the Egyptian State Railways. It will be noted that in the sketch A the sleeper is placed "on edge", whereas in the sketch B it is placed on "flat". The resistance of the sleeper against bending is greater in the first case, and consequently, the sleepers should be placed "on edge" if the distance between the axes of the rail bearers is greater than the distance between the rails (gauge). If, however, these distances are the same, the sleeper is not subjected to any bending stresses at all, and there is therefore no apparent reason to prefer the first arrangement to the second one, in this case.[1] This however, is only a desideratum, which is not adhered to in every case of practice.

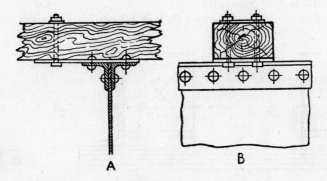

Fig. 193.

Two other devices for fixing the sleeper on the rail bearer are shown in Figs. 194 and 195. They are characteristic of modern tendencies, allowing for independent elastic deformation of sleeper and bearer.

Experiments made in America show that a wide spacing of rail bearers causes the sleepers

[1] Some specifications require a check on the resistance of sleepers for accidents, when the wheel leaves the rail, and is supported by the sleepers (directly, or through a special "counter-rail").

to afford a cushioning effect, which reduces the impact (Report of the A.R.E.A. on impact tests).

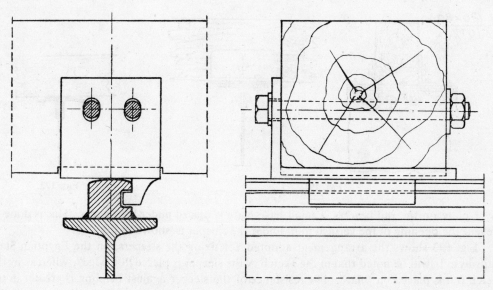

Fig. 194.

Owing to the heavy vibrations of rail bearers during the passage of a train, they should be made particularly strong and rigid. Accordingly, if the plate-girder type is adopted for rail bearers, the depth-to-span ratio is given rather a high value, preferably, one-sixth to one-eighth, and at any rate, not less than one-tenth.

It is stipulated in some specifications that in calculating the pitch of the rivets in rail

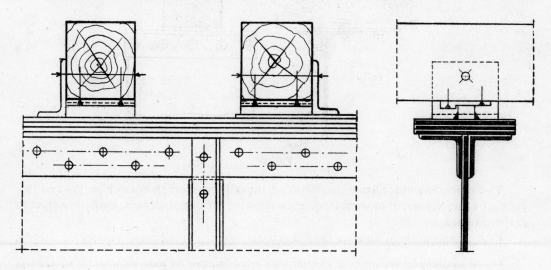

Fig. 195.

PLATE-GIRDER BRIDGES

bearers, the vertical load directly applied to the top flange must also be taken into account. According to formula (A), p. 221, the stress per rivet was

$$t = \frac{QS_f p}{I}.$$

This stress was imagined to be applied horizontally; to this we must now add a vertical force. In calculating the latter we shall assume that the load of a wheel of an engine is distributed by the rail over three sleepers. Each rivet will then take $v = P\dfrac{p}{2a}$, where a is the spacing of the sleepers. Resolving the two stresses into one resultant, we obtain (per rivet):

$$\sqrt{\left(\frac{QS_f p}{I}\right)^2 + \left(P\frac{p}{2a}\right)^2} = p\sqrt{\left(\frac{QS_f}{I}\right)^2 + \left(\frac{P}{2a}\right)^2} = pB.$$

Wherefrom

$$f_s \geqslant \frac{pB}{\dfrac{\pi d^2}{2}} \qquad\qquad p \leqslant \frac{f_s \pi d^2}{2\sqrt{\left(\dfrac{QS_f}{I}\right)^2 + \left(\dfrac{P}{2a}\right)^2}}$$

$$f_b \geqslant \frac{pB}{bd} \qquad\qquad p \leqslant \frac{f_b b d}{\sqrt{\left(\dfrac{QS_f}{I}\right)^2 + \left(\dfrac{P}{2a}\right)^2}}$$

These formulae constitute an approximate, but sufficiently precise solution. They are also valid for the main beams of the bridge, if the latter are designed as plate girders, and the sleepers are placed directly upon them (as in *Type I*, p. 214).

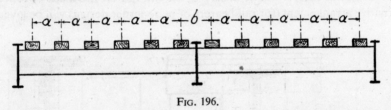

FIG. 196.

With reference to Fig. 196, all the distances a must be equal. For instance, in Egypt $a = 50$ to 55 cm. Slightly larger values are allowed on the Continent.

So far as possible b must also be equal to a, although this cannot always be satisfied.

It is also clear that in order to obtain a satisfactory spacing of rivets, the pitch, p, must be equal to a/n, where n is an integer.

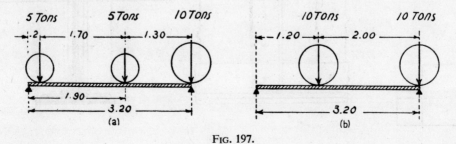

FIG. 197.

On the other hand as a general rule, p must not be greater than 10 cm (4 inches).

The rivets connecting the rail bearer to the cross-girder are designed according to the maximum shear, as is shown in the following example:

Example. Find the maximum shear for the rail bearer mentioned in the foregoing example (see pp. 225 and 226).

(1) *Case (a)*. One large and 2 small wheels on the rail bearer (see Fig. 197 (a)). The reaction is:

$$R = \frac{10 \times 3.2 + 5 \times 1.9 + 5 \times 0.2}{3.2} = 13.28 \text{ tons.}$$

(2) *Case (b)*. Two large wheels on the span (see Fig. 197 (b)):

$$R = \frac{10 \times 3.2 + 10 \times 1.2}{3.2} = 13.75 \text{ tons.}$$

The maximum value of the shear, evidently, corresponds to the second case. This case will, therefore, be included in the design of the connection.

The shear due to the structural weight of the sleepers is always small.

If p_1 is the weight of the track, per unit length, and p_2 is the weight of the rail bearer, this maximum shear is equal to

$$\left(\frac{p_1}{2} + p_2\right)\frac{l}{2} = \frac{w}{2}.$$

A structural example illustrating the method of connecting rail bearers to cross-girders is given in Fig. 198. In this case, it will be observed, the rail bearers are made of rolled joists.

An alternative arrangement (for rail bearers of plate-girder design) will be given later on.

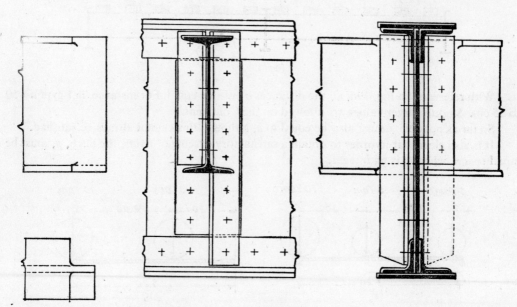

Fig. 198.

PLATE-GIRDER BRIDGES

2. Design of Cross-girders

The purpose and function of cross-girders is twofold: (*a*) to support the rail bearers and (*b*) to keep the main girders in a vertical position.

Numerical calculations are usually made with reference to the first objective alone, but the second point must also be borne in mind, when detailing the structural design.

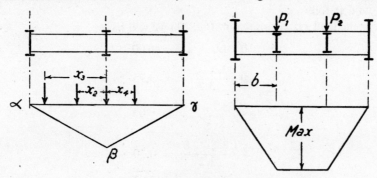

Fig. 199.

Calculations. The main factor is the live load, for the dead load is, in this case also, relatively small.

The live load is not applied directly to the cross-girder, but is transmitted through rail bearers. The forces P_1 and P_2 (see Fig. 199) are therefore always equal, and are acting at the same points. The load diagram is, consequently, symmetrical.

Hence, the diagram of the live bending moment is a symmetrical trapezium, the maximum bending moment being $P \times b = M_{max}$.

It follows that in order to determine the critical position of the loads, it suffices to find the maximum values of the loads P. This is the sum of the reactions of two consecutive rail bearers.

The effect of a load A at a distance x from the cross-girder is

$$A \frac{l-x}{l}.$$

Should there be a series of loads A_1, A_2, A_3, etc., located at x_1, x_2, x_3, etc., from the cross-girder under consideration, the value of P would be found as follows:

$$P = A_1 \frac{l-x_1}{l} + A_2 \frac{l-x_2}{l} + A_3 \frac{l-x_3}{l} + \ldots = \Sigma \left(A_n \frac{l-x_n}{l} \right).$$

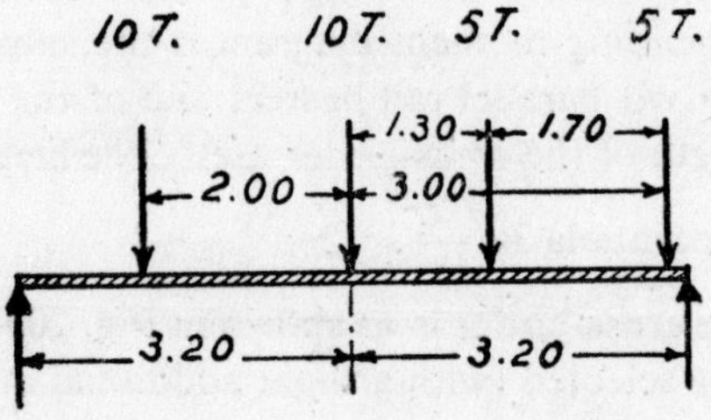

Fig. 200.

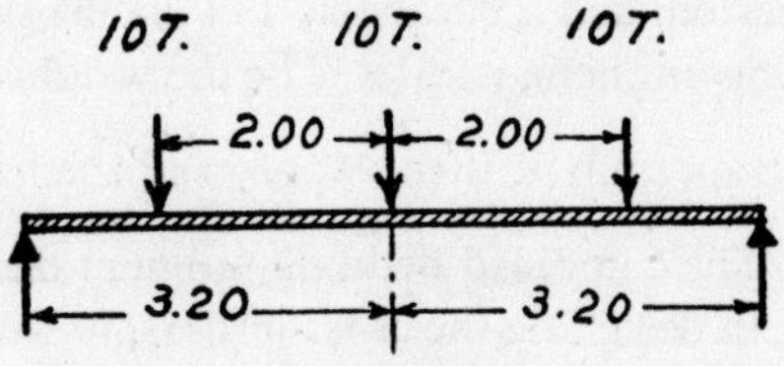

Fig. 201.

To obtain the maximum value of P, place one of the largest wheels on the girder ($x=0$) and the other wheels, as near to it, as possible. Which one of these wheels gives the *maximum maximorum* effect, must be ascertained by the trial-and-error method as shown in the following example.

Example. For the rail bearer considered on pages 225 and 230, the *maximum maximorum* value of P is found as follows:

Case I. For the load system shown in Fig. 200 we will have:

$$10 \times 1.0 = 10.00$$

$$10 \times \frac{3.2-2.0}{3.2} = 3.75$$

$$5 \times \frac{3.2 \times 1.3}{3.2} = 2.97$$

$$5 \times \frac{3.2-2.0}{3.2} = 0.30$$

$$\text{Total} \quad 17.03$$

Case II. According to the diagram in Fig. 201.

$$10 \times 1.0 = 10.00$$

$$10 \times \frac{3.2-2.0}{3.2} = 3.75$$

$$10 \times \frac{3.2-2.0}{3.2} = 3.75$$

$$\text{Total} \quad 17.05$$

The second case is, obviously, the *maximum maximorum* and must therefore be included in the calculation.

The next point to be considered is the shear.

As the live load is applied at points b and c (see Fig. 202) only, the live and impact shear is constant within the limits ab and cd. In the middle part of the beam (between b and c) the live and impact shear $=0$. The dead shear, due to the loads w (structural weight of track and bearer), is similar to the live shear, whereas the structural weight of the cross-girder itself gives a straight-line diagram.

The combined shear diagram is, therefore, as shown in Fig. 203.

Since no direct load is applied to the cross-girder, the pitch of the rivets is calculated from the formula A for horizontal shear only, as given on page 221.

The dead load for a cross-girder is the sum of its own structural weight plus the dead-load reactions of the rail bearers. Hence, the dead-load bending-moment diagram is the sum of a trapezium and a parabola. Let $2W$ be the weight of two parallel rail bearers and of the track resting on them, and let C be the weight per unit length of the cross-girder itself. The height of the trapezium is, then, $W \times b$, and the height of the parabola is $\frac{CL^2}{8}$.

The combined bending-moment diagram for the cross-girder is as shown in Fig. 203.

In designing the cross-girders, its section must be selected (with a slight additional margin of safety) depending on the value of

PLATE-GIRDER BRIDGES

$$P \times b \times (1+k) + \frac{W}{2} \times b$$

in which the letter k represents an appropriate impact allowance.

The value $\frac{CL^2}{8}$ is then calculated, and the length of the cover plates is determined from the combined bending-moment diagram, inclusive of all loads and forces.

Structural Details.—The structural details of a cross-girder are illustrated in Fig. 204. It has already been intimated that in addition to supporting the rail bearers, a cross-girder must, also, be capable of maintaining the main girders in their true, vertical position, which is particularly important in all trough bridges. With this object in view, the connection of the cross-girder with the main girders is given particularly great rigidity, and the number of the rivets in this connection is made larger than that required by the usual stress calculation alone. *An especially significant contingency*, to be taken into account in this part of the design, is that the upper flange of the main girder, if it is not properly supported, may eventually buckle. In this connection, it will be realised, that the objective of the heavy gusset *abcd* (see Fig. 204) is to provide sufficient rigidity against such buckling.

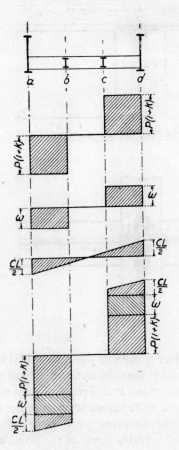

Fig. 202.

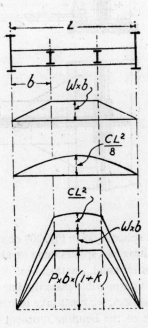

Fig. 203.

Such a contingency is less significant with the typically English, broad-flanged design shown in Fig. 184 (a), because this type is less likely to buckle. It is therefore only natural that the heavy triangular gusset, shown in Fig. 204, is frequently dispensed with in English designs. On the other hand, in Fig. 205, which shows an American cross-girder, the triangular gusset in question is not only present, but in addition it is reinforced with angle-bars at the edge, because American plate girders are particularly deep, and require therefore more stiffening.

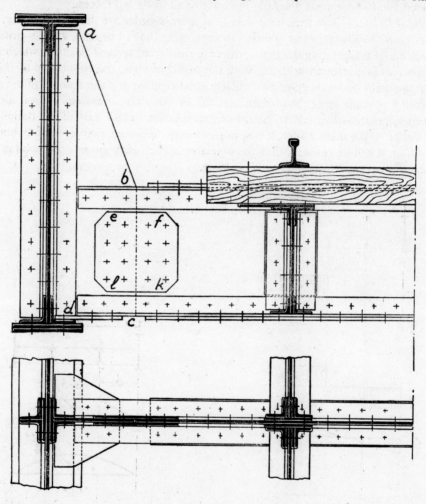

Fig. 204.

The splices *efkl* in Fig. 204 constitute the cover plates of the web joint. The total sectional area of these two splices must be at least 20 per cent greater than the sectional area of the web. On the other hand, the thickness of each splice must not be less than $\frac{3}{8}$ inch, because this is the minimum permissible plate thickness in a railway bridge. With regard to the number of rivets in these splices, we must take into account that, apart from the primary stress due to the shear, the effect of the bending moment causes an additional shearing stress in the rivet, which is supposed to be proportional to the distance between this rivet and the centroid of the group.

PLATE-GIRDER BRIDGES

In fact, owing to the elasticity of the metal, the plate $abdc$ (see Fig. 206) yields under the action of the moment Px, and the elastic movements are then proportional to the distances $n_1, n_2, n_3 \ldots$

This must be true for the shearing stresses also. We thus obtain:

$$f_1 = Cn_1 \qquad f_2 = Cn_2 \qquad f_3 = Cn_3 \ldots$$

in which C is a constant.

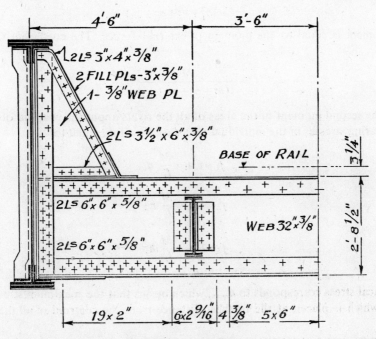

Fig. 205.—Half-section.

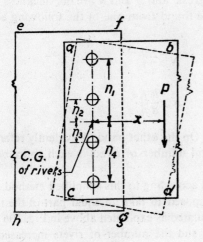

Fig. 206.

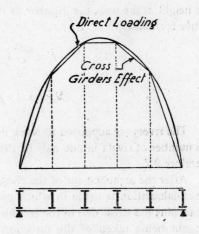

Fig. 207.

Hence, the total forces taken by the individual rivets with sectional areas a are

$$v_1 = f_1 a = aCn_1 \qquad v_2 = f_2 a = aCn_2 \qquad v_3 = f_3 a = aCn_3 \qquad \ldots$$

The moment of the internal forces is the sum of the moments of all the rivets, namely,

$$av_1 n_1 + av_2 n_2 + av_3 n_3 + \ldots$$
$$= aCn_1^2 + aCn_2^2 + aCn_3^2 + \ldots$$
$$= Ca\,(n_1^2 + n_2^2 + n_3^2 + \ldots)$$

This moment is equal to the moment of external forces. The coefficient C is therefore equal to:

$$C = \frac{M}{a\,(n_1^2 + n_2^2 + n_3^2 + \ldots)} = \frac{M}{I_r}$$

where I_r is the second moment of the areas of all the rivets about the centroid of the group.

The shearing stresses in the individual rivets are therefore equal to:

$$f_1 = Cn_1 = \frac{M}{I_r}\,n_1$$

$$f_2 = Cn_2 = \frac{M}{I_r}\,n_2$$

$$f_3 = Cn_3 = \frac{M}{I_r}\,n_3$$

$$\ldots$$

The critical stress corresponds to n_{max}, which means that the maximum stress takes place in the rivet which is placed at the furthest distance from the centroid of all the rivets, taken together.

In designing such a cover plate in practice, we begin by assuming that the number of rivets must be sufficient to transmit the maximum shear which the web plate can possibly carry.

In other words, if f_s is the permissible shearing stress, and b and h are the thickness and the height of the web, the number of rivets, N, will be found from one of the following alternative formulae:

$$N \times \frac{\pi d^2}{2} \times f_s \geqslant bhf_s \qquad \text{or} \qquad N \geqslant \frac{2bh}{\pi d^2}$$

$$N \times b \times d \times f_b \geqslant bhf_s \qquad \text{or} \qquad N \geqslant \frac{h}{d}\frac{f_s}{f_b}.$$

The rivets are supposed to work in double shear. On the other hand, N evidently refers to the number of rivets in one half of the splice, the total number of rivets in each splice being therefore $2N$.

After the arrangement of the rivets is established according to this simplified method, another calculation is made in which every rivet is supposed to take an equal part of the total shear, plus the stress due to the bending moment calculated as explained above in Fig. 206 due account being taken of the direction of this stress, and the number of rivets increased, if necessary.

PLATE-GIRDER BRIDGES

The number of rivets in the vertical angle-bars connecting the cross-girder to the main girder, is usually much greater than required to resist the reaction of the cross-girder.

SECTION IV. CALCULATION OF THE MAIN GIRDERS

1. Standard Type

Main beams of the plate-girder type are designed according to the bending-moment diagram. If the sleepers rest directly on the upper flange of the girder, this diagram is practically a curve. For bridges with cross-girders, the diagram is a polygon.

Assuming the load diagram and the span to the same in both cases, the apices of the polygon of the second type, fall on the curve representing the bending moment for the first type (see Fig. 207).

Hence, the curved diagram prepared on the assumption of a directly applied load may be used, as a first approximation, in preparing the bending-moment diagram for the second case.

A bending-moment diagram may be computed either arithmetically, or graphically. In the case of railway bridges, graphical methods are frequently preferred, for they are believed to be more expedient, in the case of a large number of wheels in the load diagram. Two graphical methods are most often used: the influence-line method and Mohr's funicular polygon.

The influence line AxB for the bending moment at the section y (see Fig. 208) is obtained by joining A to D and B to C, AC and BD being respectively equal to Ay and By. In this method the vertical scale for moments is the same as the horizontal scale for lengths. The maximum value of the live moment is calculated by placing one of the largest wheels over the section considered (that is, y, in our case) and measuring, on the drawing, the ordinates corresponding to the various wheels on the span. The area of the triangle AxB, multiplied by the weight of the main girder (per unit of its length) gives the moment due to its structural weight. The ordinates of the influence line corresponding to the positions of the cross-girders, multiplied by the dead-weight reactions of each cross-girder, give the corresponding bending moment. An influence line, such as that described above, is easily drawn for each panel point.

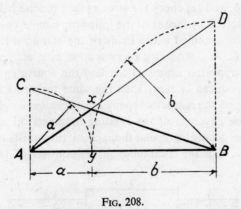

FIG. 208.

On the other hand, Mohr's method consists in drawing a funicular polygon for the set of moving loads representing the train, and then placing the span in various positions under these loads (see Fig. 209).

The details of the procedure are given in all technical text-books and engineering encyclopaediae.

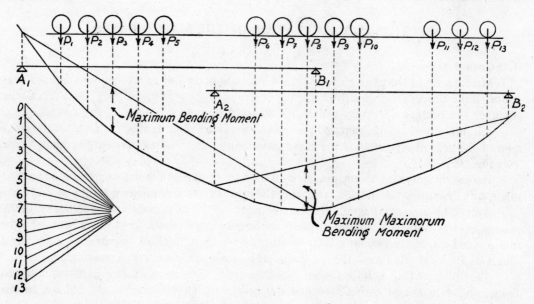

Fig. 209.

2. Continuous Main Girder

This type is of particular interest in connection with swing bridges, which are usually of continuous design.

A continuous girder is a statically indeterminate system, and consequently, the exact stresses in such a girder cannot be determined, until the girder itself is designed.

It follows that such a calculation is always divided into two parts: (1) design the girder by some approximate method, and (2) check the stresses by a precise calculation.

The problem is best solved by means of the following simple computation:

In accordance with the specified wheel-loads of the standard locomotive, find the maximum bending moments M_a, M_b, M_c,... for every span a, b, c, etc., assuming that these spans are independent, simply supported beams. Then find the equivalent distributed loads, p_a, p_b, p_c,... which would have produced in every span the same moments M_a, M_b, M_c,... Assuming then, that the loads acting on the respective spans of the *continuous* girder are indeed p_a, p_b, p_c, etc. (see Fig. 210), draw the envelope of the bending moments on this conventional assumption, design the plate girder accordingly, and then, check the results by the precise method.

Suppose, for instance, that the continuous girder consists of two equal spans, L, each.

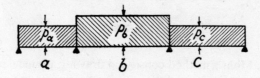

Fig. 210.

PLATE-GIRDER BRIDGES

First, find the maximum moment for a simply supported girder of the span L. Let the maximum live bending moment for the span L be M'_e and let the dead load per unit length of girder be p_d.

The conventional continuous live load p_e is found as follows:

$$M'_e = \frac{p'_e L^2}{8} \qquad \therefore p'_e = \frac{8M'_e}{L^2}.$$

Consider two alternative cases: (a) both spans loaded, (b) one span loaded and the other one free.

Case (a). The maximum negative moment is found from the equation of three moments:

$$M_1 L_1 + 2M_2(L_1 + L_2) + M_3 L_2 = -\tfrac{1}{4}(p_1 L_1^3 + p_2 L_2^3).$$

In our case:
$$M_1 = M_3 = 0$$
$$L_1 = L_2 = L$$
$$p_1 = p_2 = p'_e + p_d,$$

consequently
$$4M_2 L = -\tfrac{1}{2}[p'_e + p_d]L^3$$

and it follows that
$$M_2 = -\frac{(p'_e + p_d)L^2}{8}.$$

The computation of the bending-moment diagram in this case is done as follows (see Fig. 211): draw two symmetical parabolas (one for either span), the maximum ordinate being in both cases $= \frac{L^2}{8}(p'_e + p_d)$. Then, trace the triangle abc with the maximum ordinate equal to M_2

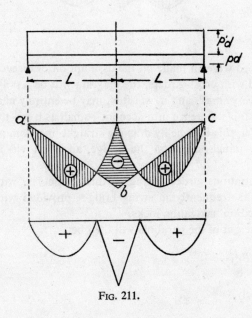

Fig. 211.

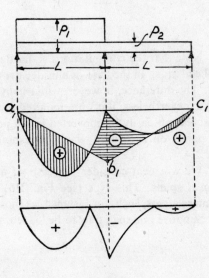

Fig. 212.

and show in a new diagram the differences between the ordinates of the parabola and those of the triangle.

Case (*b*). As in the first case: $M_1=M_3=0$ and $L_1=L_2=L$, but $p_1=p'_e+p_d$ and $p_2=p_d$. Hence

$$4M_2L = \tfrac{1}{4}[p_dL^3 + p'_eL^3 + p_dL^3] = \tfrac{1}{4}[2p_d+p'_e]L^3,$$

from which
$$M_2 = -\frac{L^2}{16}[2p_d+p'_e].$$

The diagram is computed as follows: draw for the first span a symmetrical parabola with a maximum ordinate $\frac{L^2}{8}(p_d+p'_e)$, and for the second span, another parabola, with the maximum ordinate $\frac{L^2}{8} p_d$.

The height of the triangle $a_1b_1c_1$ (see Fig. 212) is equal to M_2. Then measure on the drawing the differences between the ordinates of the triangle and the parabolas and plot them in a new diagram.

In the final diagram (see Fig. 213) the curves obtained in this manner, for the first and second cases respectively, are superposed, in order to obtain the envelope of moments. This final drawing (see Fig. 213) serves for the design of the plate girder, as previously explained on page 219. It is self-evident that the diagram for the second case must be plotted twice, first, as shown, and secondly, starting from the other end.

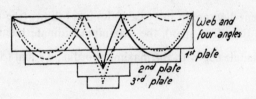

Fig. 213.

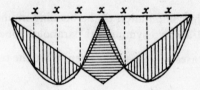

Fig. 214.

Note that the computations so far described were all made on the assumption of directly applied loads. In the case of a bridge provided with cross-girders, the diagram will be a polygon. The difference, however, will usually be very small, and most often, may be entirely neglected; nevertheless, if it were required to take this difference into account—small as it is—the parabola can easily be converted into a polygon; this is done by drawing straight lines joining the points on the curve which correspond to panel-points on the bridge, all as shown in Fig. 214.

We will next consider the case of a continuous girder resting on three supports, with *unequal* spans. This type (see Fig. 215) occurs frequently in swing bridges provided with counterweights, such, for instance, as those used on navigation locks.

Suppose one span only to be loaded: the height of the parabola will then be:

$$m_1 = \frac{p_1 l_1^2}{8}.$$

In the equation of "three moments", namely,

PLATE-GIRDER BRIDGES

we must next substitute

$$M_0 l_1 + 2M_1 (l_1+l_2) + M_2 l_2 = -\tfrac{1}{4}(p_1 l_1^3 + p_2 l_2^3)$$

$$M_0 = M_2 = 0 \quad \text{and} \quad p_2 = 0.$$

This will give

$$2M_1 (l_1+l_2) = -\frac{p_1 l_1^3}{4} = -2l_1 \left(\frac{p_1 l_1^2}{8}\right) = -2l_1 m_1,$$

from which we obtain:

$$\frac{M_1}{m_1} = -\frac{l_1}{l_1+l_2}.$$

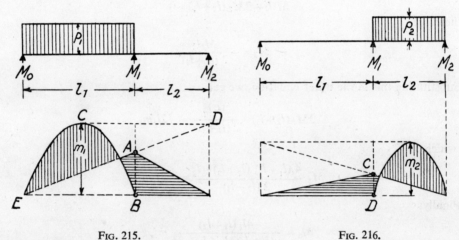

Fig. 215. Fig. 216.

It follows that the moment over the central bearing can be obtained graphically by drawing the horizontal CD (see same drawing), and then joining E with D.

Should the second span be loaded, then $m_2 = \dfrac{p_2 l_2^2}{8}$.

Consequently

$$\frac{M_1}{m_2} = -\frac{l_2}{l_1+l_2}.$$

Here again, the same graphical method can be used for finding the moment over the central support (see Fig. 216).

If the two spans are loaded, then $M_1 = \overline{AB} + \overline{CD}$ (see Fig. 217).

Note that the graphical computation described here is original, and was first published by the author in 1957, in Volume 2 of his book on *Irrigation and Hydraulic Design*, page 750.

In the case when the ratio of the width to the length of a swing span is (relatively) large—as is, for instance, the case with the road bridges for several lines of traffic—the reaction transmitted by the rollers to the girder cannot be assumed to be concentrated in one point only, but must be assimilated to the effect of two bearings; the distance between them (see l_2 in Fig. 218) being equal to the diameter of the rim of the roller path. We will then have a continuous girder on four supports, with three unequal spans. Hence, there will be two equations to solve:

$$M_0 l_1 + 2M_1 (l_1+l_2) + M_2 l_2 = -\tfrac{1}{4}(p_1 l_1^3 + p_2 l_2^3)$$

$$M_1 l_2 + 2M_2 (l_2+l_3) + M_3 l_3 = -\tfrac{1}{4}(p_2 l_2^3 + p_3 l_3^3).$$

The unknown values in these equations are the moments M_1 and M_2, whilst $M_0 = M_3 = 0$. The first solution is obtained for the case when $p_2 = p_3 = 0$, namely:

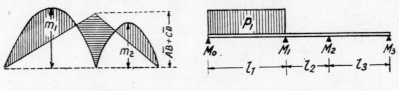

FIG. 217. FIG. 218.

$$M_1 l_2 + 2M_2 (l_2 + l_3) = 0$$

and therefore

$$M_2 = -\frac{M_1 l_1}{2(l_2 + l_3)}.$$

Substituting this in the other equation, we get

$$2M_1 (l_2 + l_3) - \frac{M_1 l_2^2}{2(l_2 + l_3)} = -2l_1 m_1$$

from which

$$M_1 \frac{4(l_1 + l_2)(l_2 + l_3) - l_2^2}{2(l_2 + l_3)} = -2l_1 m_1.$$

Finally:

$$M_1 = \frac{4l_1 (l_1 + l_2)}{4(l_1 + l_2)(l_2 + l_3) - l_2^2} m_1,$$

which yields the first solution.

The following combinations must also be considered:
1. Dead load on all the three spans.
2. Live load on central span only.
3. ,, ,, ,, extreme spans only.

When all these cases have been investigated, and envelope is drawn, in the same manner as was explained for the bridge on three supports, and the cover plates are proportioned accordingly.

3. Precise Stress Calculation for Continuous Plate-girder Bridges

After the plate girder has been designed by one of the approximate methods described in the foregoing subsection, the stresses can and must be calculated more precisely by means of influence lines, taking into account the designed sections, and replacing the conventional *distributed* load by the true *concentrated* wheel-loads.

This solution is derived from the shape of the elastic-deflection curve of the beam.

Suppose it is required to draw the influence line for the reaction at a, for the continuous girder shown in Fig. 219 A. A unit load is than placed at an arbitrary distance, x, from a, and the bearing a is supposed to be removed. The beam will then deflect as shown in the same figure, in diagram B.

Let us now apply to the left-hand end of the beam a certain force N, capable of raising

PLATE-GIRDER BRIDGES 243

this end to its original position (see Fig. 220).

It is self-evident that the stress conditions in this new cantilever beam, under the effect of the forces unity and N, are precisely the same as those of the prototype continuous girder, when it was loaded with the force unity alone, because the effect of the bearing a is now replaced by an equivalent force N. It follows that this latter force is equal to the unknown reaction of the continuous girder at a.

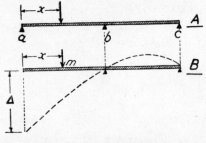

Fig. 219.

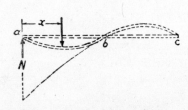

Fig. 220.

The greater the deflection Δ in Fig. 219 B, the larger must be the force N in Fig. 220. This may be transcribed as follows:

$$N = K\Delta,$$

in which K is a certain coefficient. It follows that, except for the scale, the influence lines for the deflection Δ, and for N, must be identical.

Suppose, now, we were to place the load "unity" at the left end of the beam, and remove the bearing a as before.

It may be proved that the diagram of the elastic axis for this particular case of loading also supplies the required influence line for the reaction at a.

In fact, according to Maxwell's theorem, the ordinate, y, at a distance x from a, (see Fig. 221) is exactly equal to the deflection Δ, in Fig. 219 B, with the unit load at m. The curve $AbDc$ in Fig. 221 is therefore the influence line for Δ, and also for N.

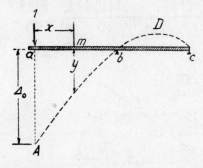

Fig. 221.

It is also obvious that with the load at a, i.e. for $x = 0$, the reaction must be equal to unity. Using then, again, the same theorem, we conclude that the coefficient $K = \dfrac{1}{\Delta_0}$. Hence

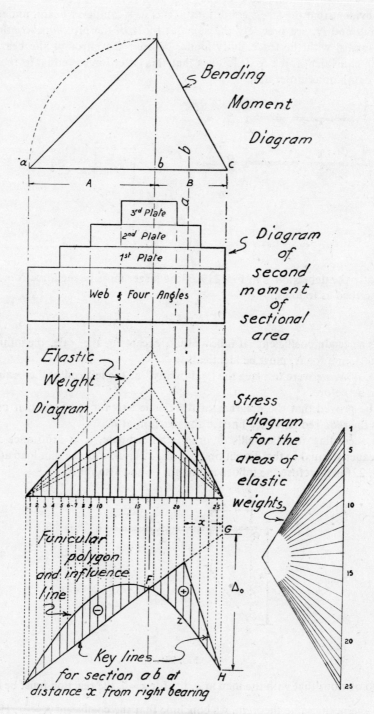

Fig. 222.

PLATE-GIRDER BRIDGES

$$N = \frac{y}{\Delta_0},$$

which determines the scale of the diagram.

It remains to be shown how this computation can be carried out, in an easy and straightforward way, with a minimum loss of time. The following graphical solution is recommended (see Fig. 222):

(a) *The bending-moment diagram.* This is a triangle, with A as maximum ordinate (see upper diagram).

(b) *The second-moment diagram*, is plotted according to the designed lengths of the flange plates (see preliminary design in the foregoing subsection). The second moments are the products of the section moduli by one half of the height of the girder, or alternatively, if the English design method is used, they are calculated as the double flange area multiplied by the square of half-the-height. The usual shape of this diagram is as illustrated by the second diagram in Fig. 222.

(c) *The elastic-weight diagram.* The ordinates of this diagram (see third sketch in Fig. 222) are the ratios of the corresponding ordinates of the two first diagrams.

(d) *The influence line for the reaction in a*, is obtained as follows: the elastic load diagram is divided into a number of vertical strips, and each strip is conventionally considered as a "force". A stress diagram is then computed for the system of forces thus obtained, and a funicular polygon is drawn in the usual manner. The points F and H of the polygon (under the bearings b and c, respectively, of the girder) are joined by a straight line, which is extended to G. This is the "key-line" for the reaction a, because the vertical distances between this line and the funicular polygon, divided by the distance Δ_0, are the "influence coefficients" for the reaction a.

(e) *The influence line for the bending moment in any arbitrarily chosen section, x—z*, is obtained by joining x with H. The influence coefficients are then the vertical distances between the funicular polygon and the broken line $HxFE$, multiplied by x and divided by Δ_0.

When the influence line is thus computed, the required bending moments are obtained by superposing the load diagram, measuring the ordinates corresponding to the wheel-positions, and summing up the products.

4. Structural Design of the Main Girders

(a) *Joints of the Web.* The spacing of the joints in the web depends upon the current sizes of plating obtainable on the market. Exceptional sizes must preferably be avoided, as the extra cost per ton of metal may often counterbalance the saving in weight and workmanship, due to a lesser number of joints. There is no universal standard for the maximum size of plating, for it varies with time and locality. In each particular case, the designer will obtain local information on the subject, before starting detailing the bridge.

In designing a joint in the web of a main plate girder, it is advisable to remember that the web forms part of the section resisting the bending moment (in case the web was taken into account in calculating the section modulus). Since at the joint of the web the section modulus is reduced, it will be necessary to provide a sufficient number of rivets in the cover plate to transmit (in addition to the shear) the bending stress from one part of the web to the next one.

Let the section modulus of the web alone be Z_w and the section modulus of the entire section (including also the web, angles and cover plates) be Z. The bending stress in the girder will then be:

$$f = M/Z_t,$$

where M is the bending moment at the joint.

The moment carried by the web alone is

$$M_w = Z_w f = M \frac{Z_w}{Z_t} = M\alpha,$$

where

$$\alpha = \frac{Z_w}{Z_t} = \frac{bh^2}{6Z_t}.$$

The number of rivets in the cover plate must then be sufficient to resist: (1) The total shear Q, (2) the moment $M\alpha$.

The shear Q is supposed to be almost evenly distributed among all these rivets. Consequently, if the number of rivets is n, the stress, per rivet, due to the shear Q, is $q = \frac{Q}{n}$.

The stress m due to the moment $M\alpha$ will be found in the same manner as was explained earlier for the joint of the cross-girder (see p. 235); namely, the greater the distance from the rivet to the centroid of the group, the larger the shearing stress produced in the rivet by the bending moment. Consequently, $m = Ky$ (see Fig. 223).

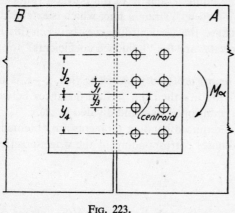

Fig. 223.

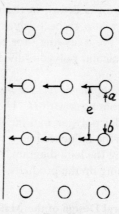

Fig. 224.

The internal moment of all the stresses acting in the rivets must be equal to the external moment $M\alpha$. It follows that

$$M\alpha = \Sigma (Ky) y = \Sigma K y^2 = K \Sigma y^2.$$

From which

$$K = \frac{M\alpha}{\Sigma y^2}.$$

Finally,

$$m = \frac{M\alpha}{\Sigma y^2} y.$$

The maximum value of m will correspond to the maximum y; consequently,

PLATE-GIRDER BRIDGES

$$m_{max} = \frac{M\alpha}{\Sigma y^2} y_{max}.$$

The total force taken by the rivets in the cover plate is the geometrical sum of all the n and m.

In this calculation the horizontal component of the distance from the centroid to the individual rivet is most often neglected (as shown in the drawing).

The vertical spacing, e, of the rivets in the plate is found as follows: the section of the plate between the points a and b (see Fig. 224) must be designed for the maximum stress which N rows of rivets may possibly carry (here N is the number of vertical rows in one half of the plate). With reference to the bearing limit, the maximum permissible force on one rivet is

$$f_b bd,$$

where b and d are respectively the thickness of the plate and the diameter of the rivet. On the other hand, the maximum stress for the section ab is then

$$(e-d)bf_t.$$

It follows that

$$Nf_b bd = (e-d)bf_t,$$

from which

$$Nf_b bd + f_t bd = ebf_t.$$

Consequently

$$e = \frac{Nf_b + f_t}{f_t} \times d.$$

(b) *Joints of Angle-bars and Cover Plates.* The joints in the angle-bars are spaced from 8 to

Fig. 225.

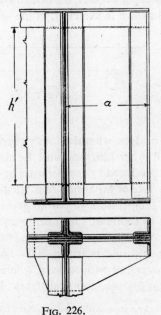

Fig. 226.

14 metres apart, according to the normal lengths of available angle-bars, or otherwise—if and as required by the erection programme.

These joints are most often covered by lag angles. It is advisable to select the lag angles of slightly smaller leg size, but of greater thickness, than the main angles, in order to have the same (or larger) area, while yet preventing the covering angle from projecting beyond the main angle, as shown in Fig. 225.

The number of rivets must, of course, be equivalent in resistance to the section of the spliced bar.

The joints of the flange plates are designed on the same principle.

(c) *Stiffeners.* Stiffeners are placed at the points of application of concentrated loads; for instance, at panel points and at the bearings.

Intermediate stiffeners are to be used if $60b < h'$, where h' is the unsupported height of the web (see Fig. 226).

The section of the end stiffener must be sufficient to carry the entire reaction of the bridge (in terms of the permissible bearing stress).

The distance from the end stiffener (which is always located over the bearing) to the end of the girder (*a*, in Fig. 226) must be sufficient to secure the bearing plate properly, and must never be less than 40 cm (16 inches) for railway bridges.

SECTION V. SECONDARY DETAILS: BEARINGS, EXTREME SLEEPERS, BRACINGS, ETC.

1. Bearings

In designing the bearings of a beam bridge, provision must be made for the angular elastic movement of the girder. Had the bearings been made flat, as shown in sketch I in Fig. 227, the edges *a* and *b* would have been overstressed and crushed. To ensure a more uniform distribution of bearing stresses, the bridge must be supported at either end on circular or parabolic surfaces, as shown in sketch II, in the same figure.

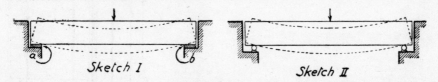

FIG. 227.

All bridges, whether simply supported or continuous, are provided with one fixed bearing only. All other bearings must be movable, i.e. they must be designed to allow horizontal expansion produced by temperature or elastic deformation (see Fig. 228).

The rate of expansion of a bridge evidently depends upon the details of the design, on the range of temperature variations and on many other factors; but in practice, it may be roughly assumed to be about $\frac{1}{8}$ inch for every 10 feet of the length of the span (in actually existing steel bridges) or about one-thousandth of the span.[1]

The movable bearings in bridges greater than 20 metres in length, are provided with turned rollers, or with rockers, whereas for bridges below 12 metres the bearings are arranged to slide on smooth, planed surfaces. For intermediate spans, i.e. between 12 and 20 metres, the

---

[1] This is a frequently used standard coefficient employed in bearing design.

PLATE-GIRDER BRIDGES

practice varies; in fact, sliding bearings may be used for such spans in the case of light roadway bridges, whilst heavy railway bridges of corresponding lengths must be provided with roller bearings, because the friction force evidently depends on the magnitude of the reaction.

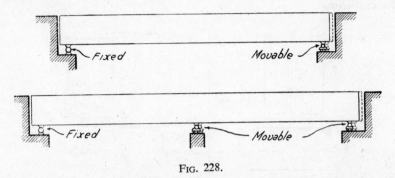

FIG. 228.

Schaper recommends sliding bearings if the reaction due to structural weight does not exceed 12 tons; beyond this limit the bearings should be provided with rollers or rockers.

The sketch in Fig. 229 shows a sliding bearing for a plate-girder bridge. The bed-plate, which is here made of cast iron, or cast steel, is placed on the bearing stone, with a layer of cement about $\frac{3}{8}$ inch thick in between metal and stone. The bed-plate is erected in position immediately upon the casting of the concrete and before the latter has set, so as to provide as close a contact as possible, between metal and cement. In older designs, the bed-plates were fixed to the masonry by means of bolts, but, according to more recent practice, the same object is achieved by ribs cast on the under surface of the bed-plate.

The area of the bed-plate must be sufficient to keep the bearing stress on the stone within the permissible limit. With reference to the sketch in Fig. 229, we may write

$$\Theta \geqslant \frac{A}{Ca}$$

where A is the total maximum reaction,
  Θ is the permissible bearing stress for ashlar, viz. about 30 kg/cm² (but for best kind of ashlar the maximum is 40 kg/cm²).

Consequently, $Ca \geqslant \dfrac{A}{30}$ if A is expressed in kilograms. The width, a, of the bed-plate is usually made about 1.25 to 1.5 times the width of the flange of the plate girder e. Assuming a mean proportion of $a = 1.35e$ we get

$$C \times 1.35e \geqslant \frac{A}{30} \quad \text{or, finally} \quad C \geqslant \frac{A}{1.35 \times 30 \times e} \sim \frac{A}{40e}.$$

Knowing the permissible bending stress, f, the thickness, b, of the bed-plate is determined as follows: assuming a uniform distribution of stress on the under surface of the bed-plate, the bending moment in the middle is $A/2 \times C/4 = AC/8$ (see Fig. 230). Hence, the bending stress is

$$f = \frac{AC}{8 \dfrac{b^2 a}{6}} = \tfrac{3}{4} \frac{AC}{b^2 a}.$$

It follows that $b^2 \geqslant \frac{3}{4} \frac{AC}{af_t}$ and $b \geqslant \sqrt{\frac{3}{4} \frac{AC}{af_t}}$.

The permissible bending stress may be taken to be 1,000 kg/cm² for cast steel and about 280 kg/cm² for cast iron.

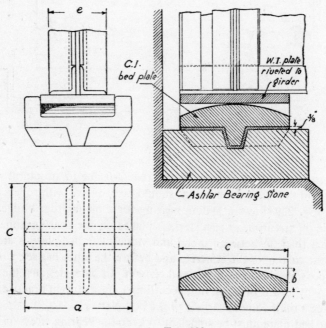

Fig. 229.

The proportions of the rib are given in Fig. 231.

The section of the rib usually takes the form of a trapezium, with about 1 to 8 side slopes.

The fixed bearings are provided with various types of "tongue and groove" arrangements, one of which is illustrated in Fig. 232. Note that whilst preventing the sliding movement of the girder, such a device must not interfere with its angular elastic deformation.

Fig. 230.   Fig. 231.

Roller bearings may, broadly speaking, be divided into two types: for spans greater than about 25 metres the arrangement includes two or more rollers, and a hinge, as shown schematically in the upper diagram in Fig. 233; whereas for spans less than 25 metres one single roller will usually suffice, and the hinge will in this case (see lower diagram in Fig. 233) be dispensed with. It will be realized that the critical span of 25 metres quoted above is a general guide only.

PLATE-GIRDER BRIDGES

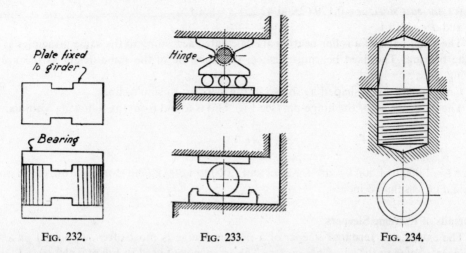

Fig. 232.   Fig. 233.   Fig. 234.

In all cases, rollers are provided with flanges and pins (Figs. 234 and 235).

The pins are 2 to 3 cm in diameter and are designed to keep the rollers in place so long as the bridge itself does not move. The upper part of the pin is shaped either as an involute or as a cycloid (see Fig. 234), on the same principle and by the same method as a tooth of a gear-wheel. The diameter of a bridge roller should never be less than 3 inches (about 8 cm).

The formula for calculating the diameter of the rollers is as follows (Herz' formula):

$$f = 0.42 \sqrt[2]{\frac{AE}{ar}},$$

from which we obtain $ar = 0.179 \dfrac{EA}{f^2}$ (see Fig. 235). The permissible stress f is much greater in this case than the critical limits adopted in other calculations, namely:

   4,000 kg/cm² for cast iron
   5,000 „   „ wrought iron
   6,500 „   „ cast steel

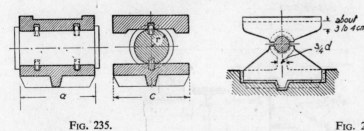

Fig. 235.   Fig. 236.

From this we obtain:
(a) *for cast iron:* $ar = 0.179\,(1{,}000/16)\,A = 11.2 A$
and $d \times a = 23 A$, where A is in tons, and d and a are in centimetres.
(b) *for wrought iron:* $ar = 0.179\,(2{,}150/25)\,A = 15.4 A$
and $d \times a = 31 A$.

(c) *for cast steel:* $ar = 0.179 \,(2{,}200/42.25)\, A = 9.3 A$
and $d \times a = 19 A$

The other parts of a roller bearing are calculated according to the same principles as for a sliding bearing. The fixed bearing must, in this case, be of the same height as the movable bearing.

Consequently, it is shaped as shown in Fig. 236, or on similar lines.

The diameter, d, of the hinge-pin (see Fig. 236) is found from the following formula:

$$d = 0.8 \, \frac{A}{a f_2}$$

where $f_2 = 1{,}000$ to $1{,}500$ kg/cm² for steel and $1{,}000$ to $1{,}300$ kg/cm² for cast iron, but in no case should d be less than 3 inches.

2. Details of Extreme Sleepers

The extreme or terminal sleeper of a railway bridge is most often supported on a short cantilever, riveted to the end cross-girder. The arrangement used in Egypt is shown in Fig. 237.

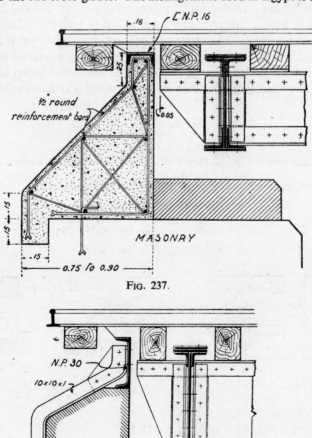

Fig. 237.

Fig. 238.

PLATE-GIRDER BRIDGES

Fig. 238 gives an alternative arrangement, frequently adopted on the European continent.

Fig. 239 shows the end sleeper of a deck bridge in which the sleepers rest directly on the main girders.

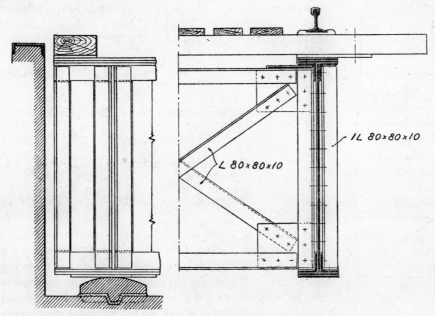

FIG. 239.

3. Wind Bracings

The calculation of wind bracings is made on the assumption of a conventional wind pressure, equivalent to a static pressure $p = 250$ kg/sq metre of the exposed surface, if the bridge is unloaded, and $p = 150$ kg/sq metre for a loaded span.

In calculating the wind loads, the train may be assimilated to a rectangle 3 metres in height, placed above the top of the rail (see Fig. 240).

The pressure of the wind causes the load on the rail a to increase, but partly relieves the load on the rail b. Consequently, the bottom flange, c, of the leeward girder (see Fig. 240) becomes overstressed. This effect is still further accentuated owing to the fact that the flange in question forms the tension chord of the horizontal lattice frame, resisting the wind pressure (see plan in same figure).

Notwithstanding such theoretical considerations, practice shows that for plate-girder bridges there is no necessity to check the wind stresses in the bottom flange of the girder, for according to almost all specifications, the permissible limit for the sum of the vertical forces and wind loads, is 15 to 25 per cent greater than that for vertical forces alone; thus, for moderate-span bridges (plate girders), the additional stress due to the wind pressure will always lie within the permissible margin of the 15 per cent of the stress due to vertical loads only.

In designing the horizontal bracings of moderate-span bridges, the governing consideration will usually be the specified standard minimum size of the angle-bars ($75 \times 75 \times 9$ mm or $80 \times 80 \times 10$ mm). The calculated wind stress in such angle-bars is always smaller than the permissible stress limit.

Take, for example, a 12-metre-span bridge and suppose that the main girders are 1.2 metres high, and are spaced 3.5 metres apart, centre to centre. Assuming the top of the rail to be about 20 cm below the top of the girder, the maximum force in a diagonal is 4,740 kg (see calculation below) whereas the safe load for an angle-bar $75 \times 75 \times 10$ mm is (assuming 20-mm rivets) about $1,200 \times (14.1 - 2.0) \times 1.0 = 14,500$ kg, which, obviously, is amply sufficient.

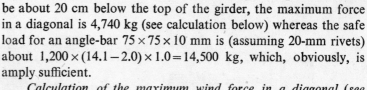

Calculation of the maximum wind force in a diagonal (see Fig. 241):

Total pressure: $12 \times 4 \times 150 = 7,200$ kg

Reaction Ra 3,600 kg

Stress in diagonal bracing $N = 3,600 \dfrac{4.61}{3.50} = 4,740$ kg

Required section, about $\dfrac{4,740}{1,200} = 3.95 \sim 4.00$ sq cm.

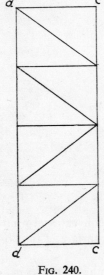

Fig. 240.

From the above calculation it must be evident that, for moderate-span bridges, the wind load cannot serve as a significant criterion for determining the required rigidity of the bridge in the horizontal direction. On the other hand, the bridge must of necessity be sufficiently rigid to withstand the shocks produced by the horizontal and vertical vibrations set up in the metal work by the locomotive. Accordingly, some specifications stipulate a horizontal live load due to vibrations of the locomotive. For instance, one of the German specifications prescribes a conventional horizontal load of six tons, acting on one of the driving wheels of the locomotive. The A.R.E.A. adopts a distributed load of 200 lb/ft (297.6 kg/metre ~ 300 kg/metre) plus one-tenth of the weight of the train, and so on.

The same result is attained in other specifications by considering a part only of the angle-bar which serves as a bracing, as efficacious (owing to the eccentricity of connections at its ends).

Hence, in practice, the standard minimum angle-bar $80 \times 80 \times 10$ mm may be adopted as a current type-section for the horizontal wind bracings, in plate-girder bridges. Three rivets, at least, must be provided at each end to secure the angle-bar in position. Short "lag" angles are often used in America to increase the resistance of these connections.

SECTION VI. OTHER TYPES OF PLATE-GIRDER BRIDGES

1. Railway Plate-girder Bridges of Other Types

In addition to the type described, many bridges are also built according to the type which we have described as *Type I* (see p. 214). This is possibly the cheapest alternative, but it is less rigid and requires much head-room between the top of the rail and the bottom of the girder. Nevertheless, under certain circumstances it may be adopted.

Let l be the span of the bridge. The minimum height of a plate girder, to prevent excessive

PLATE-GIRDER BRIDGES

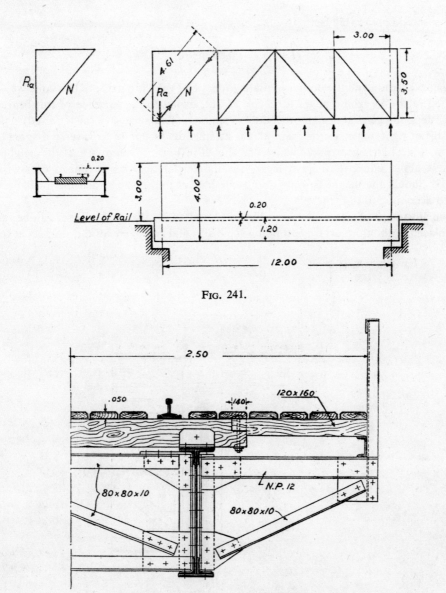

Fig. 241.

Fig. 242.—Cross-section *Type I* with side-walks.

deflection, should be at least one-twelfth (preferably one-tenth or one-eleventh) of the span. The height of the rail, plus the height of the sleeper, is about 30 cm. In addition to this, a clear distance of about 50 cm is required between the maximum water-level in the river or canal and the under side of the girder. The same margin must be provided between the car and the under side of the bridge, if the latter crosses a road. Thus the total distance between the top of the rail and the water-level or the top of the cars should be at least

$$H = \frac{1}{12} \times l + 0.50 + 0.30 \text{ metre.}$$

This should preferably be

$$H = \frac{1}{10} \times l + 0.80 \text{ metre}.$$

The calculation and the design of plate girders of *Type I* are made in the same way as was explained in the foregoing pages for *Type II* but, owing to the absence of rail bearers and cross-girders, the general design procedure is simpler. The necessary (or required) rigidity of the bridge at right angles to the main axis is attained by means of transverse diagonal cross-bracings, typical arrangements of which are shown in Figs. 242 and 243. If the height of the girders is about equal to their spacing, the diagonal braces are arranged crosswise as in Fig. 243. Should the height be only about one-half of the spacing, then the V shape will be adopted according to Fig. 242.

The advantage of side-walks as in Figs. 242 and 243 is that each sleeper can be removed and replaced, without necessarily stopping the traffic and removing all the sleepers at once.

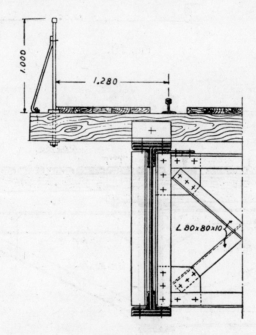

Fig. 243.

2. Plate-girder Road Bridges

Some forty or fifty years ago, plate girders were frequently used as main girders in moderate-span road bridges, but at present, reinforced concrete designs, or pre-stressed structures have almost completely replaced this type of permanent road bridge.

Broadly speaking, plate-girder road bridges are used at present chiefly as movable spans, or eventually, as approach spans to such movable spans, for the sake of continuity of design and appearance.

The variety of types of this category is much greater than it is for railway bridges. In the first instance comes the question of the flooring. On the average, three types of floors may be

PLATE-GIRDER BRIDGES

found on canal-bridges of this type. The wooden flooring was the one most frequently used in older bridges; curved steel plates occupy the second place, and reinforced concrete slabs are adopted on many modern designs.

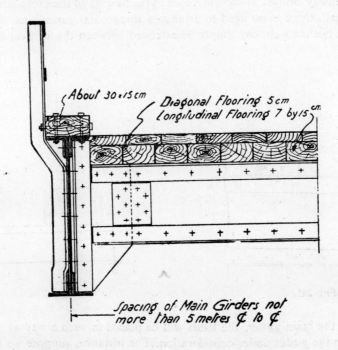

Fig. 244.

A section suitable for narrow bridges and heavy traffic is given in the sketch in Fig. 244 illustrating the timber flooring type. For bridges of greater widths it becomes more economical to support the flooring on longitudinal floor bearers, attached to the cross-girders, as in railway bridges. In this latter case, the timbers of the bottom flooring are at right angles to the main axis of the bridge.

(a) Calculations of a Timber Flooring

The upper timber flooring is generally 5 cm thick and serves only to protect the main, lower flooring from wearing out. The upper flooring is *not* included in the calculation of resistance.

The lower flooring is assumed to have a thickness of 7 to 9 cm for light traffic (6-ton cars) and 12 to 15 cm for heavier traffic (20-ton lorries and similar modern vehicles and rollers).

The permissible stress for ordinary timber (pitch-pine) is 80 to 100 kg/cm^2. The weight of the wheel is regarded as being transmitted to two planks (20 to 25 cm width each). From this, the spacing of the cross-girders is found. It should reasonably be about 1.2 to 1.5 metres. The cross-girders are designed according to the bending-moment diagram, remembering that the rolling traffic and the crowd, may assume any position on the bridge. This is the *essential difference* between the design of cross-girders in a road bridge, and a railway bridge respectively.

The connection of the cross-girder with the main girder in road bridges is usually much lighter than in a railway bridge. If the difference in the heights of the cross-girders and of the main girders is small, there is no need to arrange a trapezoidal gusset such as is shown in Figs. 204 and 244, but the web may simply be extended between the vertical angles, as illustrated in Fig. 245.

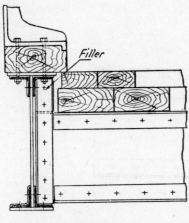

Fig. 245.

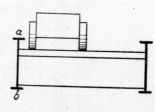

Fig. 246.

In calculating the main girder, the loads will be placed in such a way as to produce the maximum effect on the girder under consideration. For instance, suppose we have to design the girder *ab* (see Fig. 246) for a bridge which is intended to support one car only. The car must then be assumed to be placed in its nearest position to this girder, as shown on the sketch.

(b) *Curved Steel Flooring*

The steel-plate flooring, curved to form part of a cylinder, may either be suspended as shown in Fig. 247 (*a*), or it may form an arch as in Fig. 247 (*b*). The first type is more popular in Germany, the second one is preferred in France. Taking moments about the point of application of the linearly distributed load p, the tensile force H in the suspended plate (see sketch on the same chart) is found as follows:

$$wHh = p/2 \times l/2 \times w$$

$$H = pl/4h.$$

The width, w, of the plating over which a concentrated load, P, is distributed, cannot be estimated exactly from elementary consideration. Winkler gives the following formula

$$P = \frac{1{,}120h\delta - 0.018ql^2}{h + 2.4\delta} \frac{\delta}{l},$$

where P is the concentrated load in tons;
    q is the distributed dead weight in tons per square metre,
    h is the rise in cm,

PLATE-GIRDER BRIDGES

δ is the thickness of the plate in cm,
l is the span in cm.

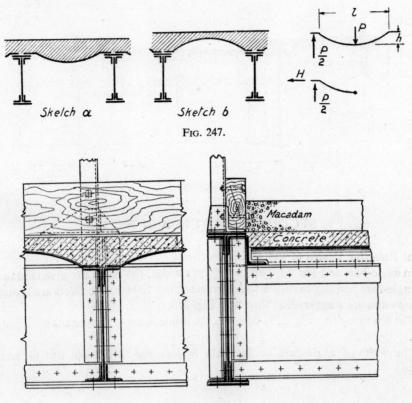

Fig. 247.

Fig. 248.

This formula is believed to correspond to a permissible stress of 800 kg/cm^2. The inverted or arched cylindrical plate must be made about 20 per cent thicker than the suspended plate to ensure elastic rigidity, i.e. resist against buckling.

Fig. 248 shows the details of a plate-girder bridge with such a cylindrical plating. It will be noted that a layer of poor concrete is placed over the plates. It serves to distribute the pressure and stiffens the cylinder.

(c) Reinforced Concrete Slab

The reinforced concrete slab of a plate-girder bridge is designed according to the same rules and methods as for rolled-joist spans, and therefore does not require to be discussed here again. A somewhat different type is shown, however, in Fig. 249, which resembles, in principle, the arrangement illustrated in Fig. 104 (ante).

(d) Main Girders, Bearings, Bracings, etc.

These parts of the metal structure are designed in exactly the same manner as for railway bridges and do not need to be considered again.

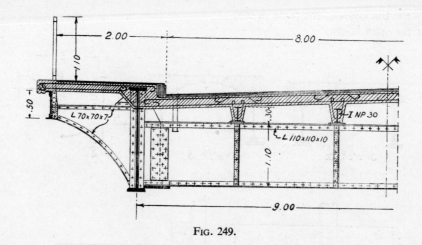

Fig. 249.

SECTION VII. RIVETS AND WELDS

1. General Rules for Design of Riveted Connections

(a) In angle-bars and flat bars up to 120 mm width, the rivets are arranged in one single line. In angle-bars and flat bars of a width greater than 120 mm the rivets are arranged in two parallel rows and are staggered, as shown in Fig. 250.

(b) The distance f (see Fig. 251) between the inside faces of rivet heads must not be greater than $4\tfrac{1}{2}d$.

(c) The width of angle-bars or flat bars W (see Fig. 252) must not be less than $3\tfrac{1}{4}d$ (better $3\tfrac{1}{2}d$).

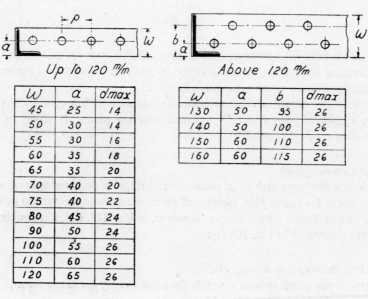

Fig. 250.

Fig. 251.

Fig. 252.

W mm	a/b	Diameter of Rivets mm 14	16	18	20	24	26
45	25	25	—	—	—	—	—
50	30	20	—	—	—	—	—
55	30	20	30	—	—	—	—
60	35	15	25	25	—	—	—
65	35	15	20	25	30	—	—
70	40	0	15	20	25	—	—
75	40	0	15	20	25	—	—
80	45	0	0	0	15	25	—
90	50	0	0	0	0	15	—
100	55	0	0	0	0	0	20
110	60	0	0	0	0	0	0
120	65	0	0	0	0	0	0
130	50/95	0	0	0	15	25	35
140	50/100	0	0	0	0	15	30
150	60/110	0	0	0	0	0	20
160	60/115	0	0	0	0	0	20

Notes

Note:

For dia of rivets d	14	16	18	20	24	26
use dia of cup D	35	40	43	45	50	55

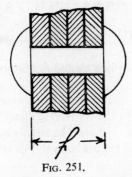

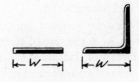

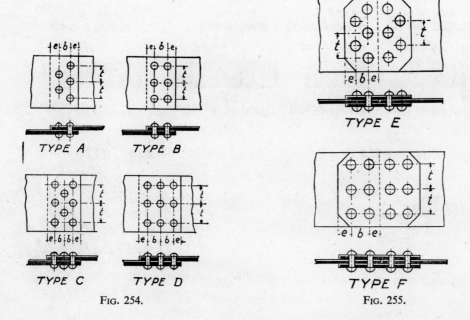

Fig. 253.

Fig. 254.

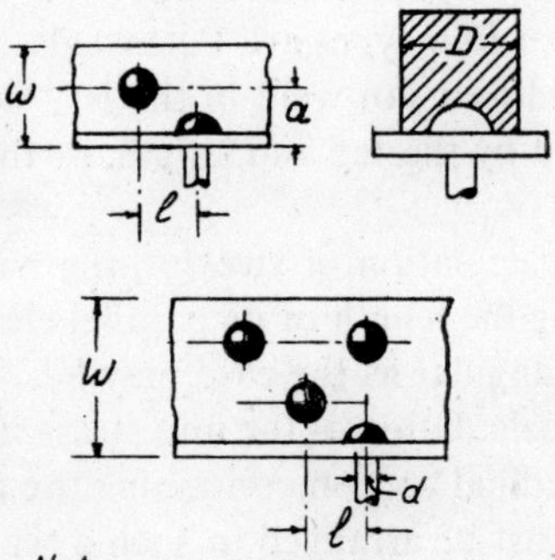

Fig. 255.

(d) The pitch of rivets in the line of stress must preferably be $3\frac{1}{2}d$ (not less than $3d$ and not more than $6d$ to $7d$, in chords).

(e) When the rivets do not carry any calculated stress, the pitch should not be greater than $8d$ (and only exceptionally $10d$).

(f) The distance between the centre of a rivet and the edge of the riveted member must not be less than $2d$ in the line of stress and not less than $1\frac{1}{2}d$ in the transverse direction.

(g) For facility of riveting the minimum dimensions of Fig. 253 must be strictly adhered to.

(h) The general arrangement of standard lap and butt joints must be as shown respectively in Figs. 254 and 255.

2. Welded Connections

Modern welding specifications include several dozens of types, which may however, all be derived from the six basic cases shown in Fig. 256.

The upper four types are butt welds, usually designated by throat dimension and length. The fillet weld and slot weld in the left-hand and right-hand bottom corners of the drawing, are designated by the leg and length, in the first case, and by the throat, depth and length, in the other.

For the calculation of stresses, the weld is considered as prismatic in form, the length of the weld being the length of its parallel elements, and the cross-section being triangular, trapezoidal or rectangular as the case may be.

For the calculation of the unit stress intensity, the critical section of every weld is assumed as the longitudinal section containing the throat of every cross-section, as shown in Fig. 256.

Welds must be arranged in such a way as to *avoid* moments about their *longitudinal* axis, and must be symmetrically disposed in relation to the loading.

If this condition is satisfied, bending stress can be transmitted through a welded joint, the section modulus being calculated for the critical section, as explained above.

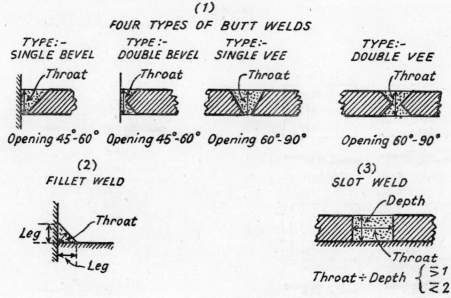

Fig. 256.

PLATE-GIRDER BRIDGES

The permissible stress for welds is usually slightly below the corresponding limits for rolled steel.

Reinforcement of welds beyond the boundaries indicated in Fig. 256 are disregarded in stress analysis.

No weld should be shorter than 8 cm (about 3 inches); the minimum leg dimension for fillet welds is 6 mm. The maximum size for such welds is twice the thickness of the thinner of the two base metals. Fillet welds carrying a specified stress must not be used singly, but must be arranged in groups of two or more, in such a manner as to avoid moments about their longitudinal axes.

In checking the stresses in welds it is always necessary to verify that the base metal of the slotted part, between the welds, is capable of carrying the calculated stress.

Chapter Eight
MOVABLE BRIDGES

SECTION I. LIFT BRIDGES

1. General Principle of a Lift Bridge

The general principle of a lift span is indeed very simple. With reference to Fig. 257, let the weight of the span from O to A, inclusive of main girders, cross-girders, deck, etc., be P_1 and let Q_1 represent the weight of the cantilever part OB, inclusive of the counterweights, etc.

The moment $P_1 \times a_1$ must be exactly equal to $Q_1 \times b_1$, for then, the centroid of the entire bridge, from B to A, will fall in the centre of rotation, O; and for any position of the span, for instance as shown by broken lines in the drawing, we shall have

$$Q_2 \times b_2 = P_2 \times a_2.$$

Note that this must be true not only in regard to horizontal distances, but also for vertical heights.

The last formula is a rather important design criterion, because the operating gear is not

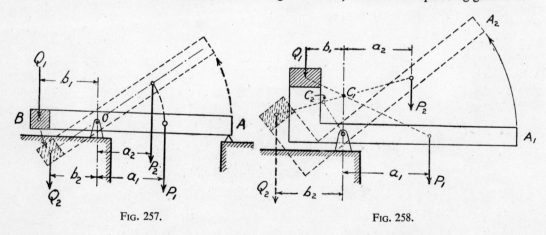

Fig. 257. Fig. 258.

intended, and *must not be allowed*, to overcome gravity forces, but is designed *for friction resistances and wind pressures only*. It is, in fact, out of the question to design a mechanical device which is strong and durable enough to overcome weight forces as an everyday routine operation; except, of course, for foot-bridges and the like.

Fig. 259.

Fig. 260.

Fig. 261.

Fig. 262.

Figs. 259–262.—The Abou El-Akdar bascule bridge.

Fig. 265.

Fig. 266.

Fig. 268.

Fig. 269.

Fig. 270.

Fig. 271.

Fig. 272.—Scherzer-type lift bridge.

2. Location of Centre of Rotation

This argument explains why it is that an arrangement such as the one shown in Fig. 258 is basically fallacious, for although the moments $Q_1 \times b_1$ and $P_1 \times a_1$ balance each other when the span is in its horizontal position, the equilibrium is immediately destroyed as soon as the bridge begins to move. In fact, although $Q_2 = Q_1$ and $P_2 = P_1$, it is nevertheless obvious that $b_2 \neq b_1$ and $a_2 \neq a_1$, which means that

$$Q_2 \times b_2 \neq P_2 \times a_2.$$

In order to make this bridge capable of being operated, the centre of rotation must be raised to C_1, and this is a rather expensive alteration, for reasons pertaining to detailed mechanical design.

On the other hand, the arrangement shown earlier in Fig. 257 is also objectionable, for the tail B of such a bridge would be below water-level when the span is open; and apart from being exposed to rust, etc., the counterweight will then lose part of its weight, owing to the effect of buoyancy. It follows that this arrangement can only be used if a watertight well is provided for housing the counterweight when the span is open, at a cost which may often exceed that of the span.

The Abou El-Akdar bascule bridge on the main railway line Benha-Zagazig, may serve to illustrate the point at issue. As seen from Figs. 259 to 261, this is a typical irrigation-canal work—but the construction of the abutment which was required to house the tail of the span (with the counterweight attached to it) when open, was a rather expensive job, as the reader will realize on inspecting Fig. 262, in which this work is represented as it could be seen before the steel superstructure was finally erected.

Apart from its generous over-all proportions, expense was also incurred in connection with erection difficulties; in fact, since this was a remodelled, old structure, serious trouble was experienced in rendering it watertight—an imperative requirement in this case, because the chamber housing the counterweight must always be kept dry. Finally, several coats of "Sika" plastering applied to the inner walls of the chamber, supplied the required solution to the problem. The remarkable point about it was that it never gave any trouble, to speak of, later on.

3. Main Shaft

According to Egyptian experience, the chief mechanical problem in attempting this type, is the main horizontal shaft, which comes out in practice to be rather too heavy, difficult to construct, and therefore, expensive.

For instance, in the case of the Abou El-Akdar bridge the main axle was almost two tons in weight and formed one of the most costly items of the estimate; notwithstanding that every step had been taken to allow to reduce its dimensions. Thus, out of the four parallel main girders of the bascule span (which are clearly seen in Fig. 262), only the outer two girders, i.e. the nearest to the bearings, are supported by the axle, and the bending moment for the latter is therefore reduced to a minimum. On the other hand, the webs of the inner two girders are provided with wide openings allowing the axle to pass through them, without transmitting any reaction. This arrangement necessitates, in turn, a special design of the diagonal counter-bracings which, when the bridge is open, carry almost all its weight, so as to relieve the axle. That is the reason they are much heavier (see Fig. 261) than might have been necessary,

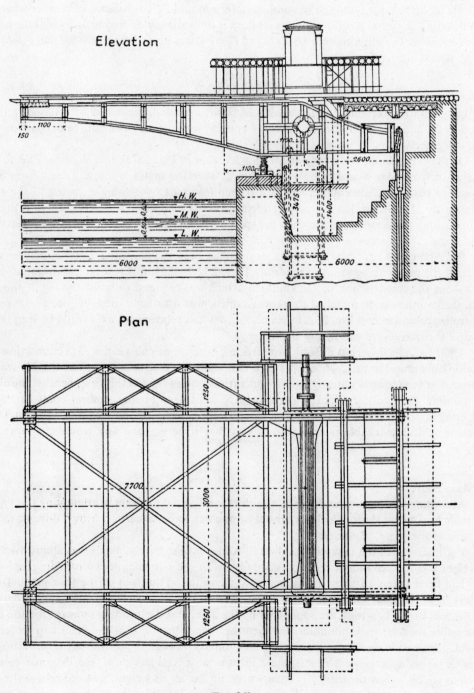

Fig. 263.

MOVABLE BRIDGES

had they been wind-bracings only (they are seen in the photograph close to the tail of the span).[1]

4. Bascule Type in Germany and America

The same kind of problem has been encountered in other countries as well, e.g., in Germany, where the bascule type has frequently been used for similar spans. In this connection attention is called to Figs. 263 and 264, which show a small bascule bridge in Rensburg.[2] In this case the axle is built of four rolled sectors and is 30 cm in inner diameter. But at the point where it crosses the web of the main girder, it widens out to 125 cm and is then replaced, beyond this point, by an expensive casting.

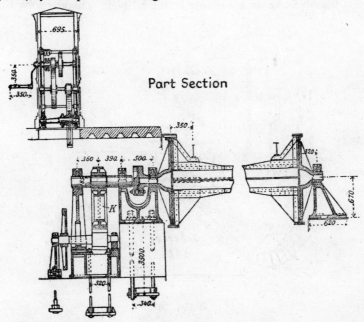

Fig. 264.

A much larger and heavier type of work based on the same (or similar) principle is generally known as the Chicago bascule bridge, but is scarcely convenient for less important canals. On the other hand, two other American types—the Strauss and Scherzer patented lift bridges—have actually been employed on the Egyptian irrigation system.

5. The Strauss Bridge

The Strauss type was used at Hamoul, on the Bagourieh Canal, in Lower Egypt. As seen from Figs. 265 and 266, which show this bridge in open and closed positions,[3] it consists of two lift spans, 13.5 metres each, with a fixed section in between, the over-all length of the whole work being 53.3 metres.

[1] The designs for this bridge were prepared by the makers, the firm Wagner-Biro A.G. of Vienna, under the control of the Bridges Service of the Egyptian State Railways.
[2] From p. 70, "Bewegliche Brücken", by W. Dietz, edited by Th. Landsberg, *Handb. d. Ingenieurwiss.*, Vol. II, Part Four.
[3] Photographs by courtesy of the makers, Cie de Fives-Lille, Paris.

Working drawings for this type are usually furnished by the owners of the patent. The following explanation is therefore confined to the main principles of the system only.

In Fig. 267 (a) let MN represent the lift span of the Strauss type and let C be its centre of rotation. The counterweight W is chosen in such a manner that the moment WR is equal to the moment Pr, P being the weight of the movable span (exclusive of the counterweight) and r the distance from its centroid, K, to its centre of rotation, C. Then, for any angle α, we may write

$$WR \cos \alpha = Pr \cos \alpha,$$

which means that in any position of the span all the forces applied to it are always in equilibrium. Since this is the objective we are aiming at, the problem is hereby solved.

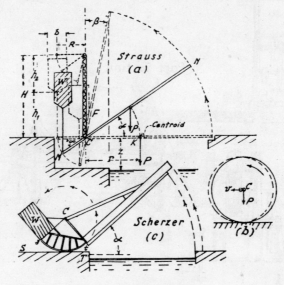

Fig. 267.

The choice of the main proportions of the bridge is based on the following considerations.

It was already mentioned earlier (see pp. 254 and 265) that in movable-bridge design, the usual tendency is to reduce to a minimum the distance z (see Fig. 267 (a)) between road-level and water-level. Since the tail of the lift span must in no case be submerged, it follows that to reduce z, we must make R as small as possible. On the other hand, the volume of the counterweight is found from

$$V = \frac{W}{\gamma} = \frac{Pr}{R\gamma}$$

in which γ is its specific weight.

It follows, therefore, that the smaller R, the lower the rail-level, but the heavier—and, consequently, the costlier—the counterweight. The engineer will have to balance the expenditure on the approaches necessary to raise the rail-level, against the extra cost of a heavier counterweight. It should be realized, however, that the latter cannot be increased indefinitely, for other reasons, also. In this connection attention is called to the tower. Its height H (see same figure) is basically the sum of: h_1, which is the height of the train or car, as the case may be,

MOVABLE BRIDGES 269

or other vehicle that must pass beneath the counterweight (inclusive of a certain margin, of course), plus h_2, which represents the counterweight itself. We cannot alter h_1 for this is one of the given conditions of the problem, but h_2 is capable of a certain range. In fact, we have

$$h_2 = \frac{V}{bw},$$

in which w is approximately equal to the width of the bridge between the main girders plus a certain constant, whilst b is the thickness of the counterweight (as shown on the drawing) capable of being selected at our convenience—within certain reasonable limits.

In fact, had b been chosen too wide, the counterweight would have struck the span in the point F and prevented it from opening completely. It is true that the span is never entirely vertical, but the tendency is to reduce the angle β to a minimum, and this is achieved by a proper choice of the main dimensions of the counterweight, b, w and h_2, and other basic parameters of the design; such for instance as R.

Turning now to the question of the main axle, attention is called to the fact that in this case, there is no continuous shaft going right through the span, but as seen from Figs. 268 and 269, two separate shafts, one on each side of the bridge. Precise adjustment in this case is a condition *sine qua non*, for these shafts must be exactly concentric. Also, rigidity is essential, to prevent eccentricity developing in the future. These, however, are obvious points, common to various other similar mechanisms.

6. The Scherzer Bridge

Consider, now, the disk in Fig. 267 (*b*). The only force resisting its movements is friction. The gravity forces, the resultant of which is applied in the centre of the disk, do not interfere, because this centre moves at right angles to the force, and the mechanical work done is therefore zero. This conclusion will still remain valid even if we attach to this disk various structural or mechanical elements, *provided their common centroid coincides with the centroid of the disk*.

This last point is the *main principle* of the Scherzer bridge (Figs. 270–272), which is supposed to be schematically represented in Fig. 267 (*c*); in fact, the counterweight W is calculated and fixed in such a manner that the centroid of *all the movable parts taken together*, falls in the centre C of the circular arc dt, and therefore, the span can roll over the rolling path ST without expanding or consuming any mechanical energy; which means that, whatever the angle α, it is always in stable equilibrium.

This type was used on the Egyptian irrigation system on the lock of the Gebel Aulia Dam (which, though located in the Sudan, was built at the expense and for the benefit of Egypt).

The bridge in question is shown in Figs. 270 to 272.[1] Particular attention is directed to the fact that in this case the main shaft is dispensed with altogether—a material mechanical advantage of the type. As already intimated, this system, as well as that of Strauss, are patented.

7. Egyptian Irrigation Type

We will next describe an arrangement which is not patented, but which is frequently used on the Egyptian irrigation canal network and elsewhere, for road bridges capable of carrying 3- to 20-ton lorries. Attention in this connection is called to Fig. 273.

Provided figure $AVFL$ is a true parallelogram, the angle α remains equal to β for all posi-

---

[1] Photographs by courtesy of Engineer Tewfik Farid.

tions of the span, which means that from the standpoint of mechanical forces, the arrangement works in exactly the same way, as if the cantilever EF were transferred in the hypothetical position KL. The ideal mechanical requirement of Fig. 257 is thus found to be realized, with the proviso however, that the tail of the bridge is not submerged when the span is open.

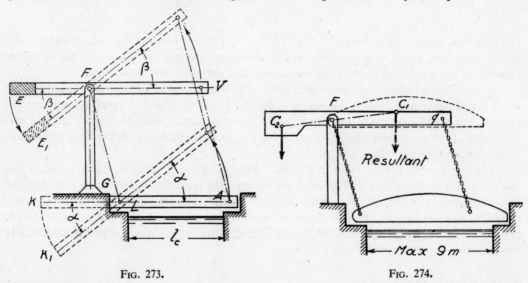

FIG. 273.   FIG. 274.

To calculate the weight of the counterweight and to find its position, we assume that the point E is transferred to the point K, while F is shifted to L, and we then proceed in the same manner as in the basic case of Fig. 257.

The principle of design incorporated in this arrangement is either used in combination with an ordinary girder, as shown in Figs. 273 to 276, or, more frequently, with a three-hinged arch as in Fig. 277.

The first type is convenient for spans up to 9 metres. It can be used for greater spans, but then, the economy of the type is rather doubtful.

The centre of gravity C_2 (see Fig. 274) of the counterweight and of the left-hand arm of the upper cantilever is found by extending the line FC_1 until it reaches the required distance from the centre of rotation. Note that in calculating the counterweight, the weight of the chain is supposed to be transferred to q.

The three-hinged design shown in Fig. 277 is rather popular in Egypt. Numbers of bridges of this type may be found on the irrigation canals of this country.

The design of the hinges recalls details of lock-gates, and bears witness to the part that hydraulic designers took in developing this type of bridge.

The three-hinged-arch frame must always constitute a separate item, apart from the floor system, as in Fig. 277. The arrangement shown in the photo, Fig. 278 is fallacious, for the compressive stress in the main girders may easily reach the breaking stress limit.

The working of the standard type is illustrated in Figs. 279 to 281.

8. Belidor or Sinusoid Type

Another type of lift span is shown in Fig. 282. It has been used on the Assiut Barrage Lock.

MOVABLE BRIDGES

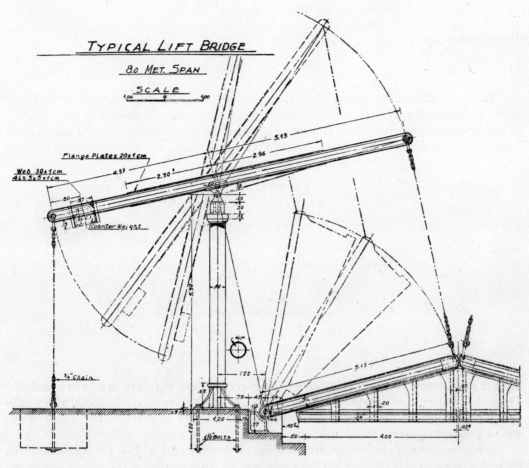

Fig. 277.

This system is referred to on the Continent either as the Belidor bridge, by the name of its first inventor, or as the Sinusoid type, according to its mathematical principle. The reason explaining the reference to the *sinus* will be clear from the following analysis. Consider diagram Fig. 283, and suppose that the lift span has risen from its original "fixed", position AB, to AD, and that, at the same time, the counterweight W, which is attached to a cable or chain, has travelled from a to b. Applying then the same principle which was already used earlier in investigating the Scherzer-type in Fig. 267 (*c*), we may conclude that the system will remain in stable equilibrium, if the total work done by the gravity forces is zero. Now, whilst the span was rising, the *negative* mechanical work done was equal to its weight P multiplied by the vertical component of the path travelled by its centroid, namely y, which is very nearly equal to $\frac{1}{2} L \sin \alpha$. On the other hand, at the same time, the *positive* work done by the counterweight was Wq. Thus, in order that the total work should be zero, or, in other words, that the centroid of all the moving parts of the system should remain on the same level, it is necessary and sufficient to satisfy the criterion

$$Wq = \frac{P}{2} L \sin \alpha$$

Hence the formula

$$q = \frac{P}{W} \times \frac{L}{2} \times \sin \alpha.$$

Thus, the *sinus* of the angle α plays an important part in the calculation of the trajectory of the counterweight, and this explains the reference to this *sinus* in describing the system. It should be realized, however, that the diagram ab is not the curve usually known as the sinusoid, for the abscissae are not, in this case, equal to the angles α, but are a more involved function

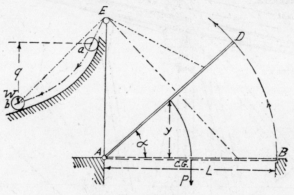

Fig. 283.

thereof; for instance, Professor Dietz shows that in certain cases the curve ab may be a cardiod.[1] The author prefers, however, the graphical method, which is more comprehensible, whilst sufficiently precise for the purposes of this investigation. To give an example, the Assiut bridge will be analysed by means of this method.

It will be observed, in the first instance, that in Assiut (see Fig. 284) the premises of the theoretical diagram Fig. 283 are not fully realized; viz. the centroid of the span is not on the same level as the centre of rotation, and the latter is not on the same vertical axis as the top pinion. Analytical formulae should therefore be amended accordingly. On the other hand, the graphical method remains unaltered.

Our first problem is the counterweight. Let P be the total weight of the movable span, and let Y be the vertical component of the entire trajectory of its centroid, i.e. the difference in levels between the uppermost and lowermost positions of this centroid. Also, let Q be the difference in the levels of the uppermost and lowermost positions of the counterweight (see Fig. 284). The principle of zero work will yield, then, for the size of the counterweight the equation

$$V = \frac{W}{\gamma} = \frac{PY}{\gamma Q}.$$

where γ is the specific weight of the counterweight.

It is obvious from this simple formula that the distance Q, on which depends the height of the tower, is inversely proportional to the volume V, which determines the cost of the coun-

[1] *Ut supra*, p. 40.

Fig. 278

Fig. 279

Fig. 280.

Fig. 281.

Fig. 282.

MOVABLE BRIDGES

terweight and all the operating machinery. The designing engineer may therefore choose these basic parameters of design, in such a manner as to obtain a minimum over-all cost.

His choice, however, is not entirely free, for in order to avoid an awkward general layout, the distance r, from the centre of rotation to the point where the cable is attached to the girder, must, on the one hand, be a certain function of the span, and on the other hand, be approximately equal to the height of the tower. Thus, ultimately, the height of the tower is correlated with the span, although the correlation is not a rigid one. As a first approximation r may be taken from $0.8L$ to $0.9L$.

Once the ratio $\xi = W/P$ is finally selected, it controls the layout of the entire bridge. In particular, the trajectory and rolling path of the counterweight are calculated point by point, as follows: assume the span to be in any position CD, and measure on the drawing the height y of the centroid F above its original level (line I–I). Multiply y by $1/\xi$ and plot the product $q = y/\xi$ downwards from the basic horizontal line o–o drawn through the upper position, Z, of the counterweight. We thus obtain the line x–x. Next, measure the length nm of the cable, between the pinion and the span, and the length Δ of the arc sn, and calculate

$$f = \text{total length of cable} - \text{length } nm - \Delta.$$

Finally, the intersection of the line x–x with the circle of radius f and centre s, yields the point z of the trajectory. Of course, f and Δ are interdependent, and the solution is therefore a trial-and-error method, but in practice, this contingency is almost negligible. In fact, the precision of the result depends chiefly on the scale adopted for the graphical computation and on the accuracy of the drawing. On the other hand, with the graphical method, it is easy to introduce various corrective factors, such as the elastic extension of the cables (which varies with the angle of inclination), the effect of the temperature, etc.

Wind reaction is the chief force to be considered in assessing the required capacity of the operating machinery. Also, as no positive drive is provided, weight alone must be capable of closing the span against the wind. Thus, if N is the force of the wind and ρ the height of the centroid of the exposed area measured above the centre of rotation c, then the minimum eccentricity e (see Fig. 284) is found from the equation

$$e = \frac{N\rho}{P}.$$

A buffer is, therefore, provided in E, thus ensuring that the eccentricity will never be less than the minimum e.

Reports from French Indo-China, where this type of bridge was used in relatively modern times, were rather unsatisfactory, but in Egypt, the Assiut Barrage and Ismailia Regulator bridges, which are both designed according to this principle,[1] appear to prove that once properly erected and adjusted, they give no trouble whatever.

So far as the author knows, the system has never been patented.

SECTION II. SWING BRIDGES

1. Two Types: Pivot-bearing and Rim-bearing

These bridges are divided essentially into two types: the pivot-bearing and the rim-bearing

---

[1] The credit for these designs is entirely due to Messrs. Coode, Wilson, Mitchell and Vaughan-Lee, Consulting Engineers, Westminster, London. The method of the foregoing analysis is the author's.

274   ARCHES AND BRIDGES

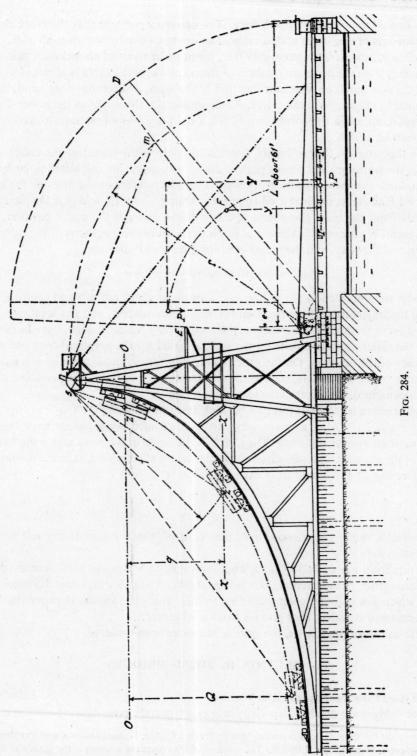

Fig. 284.

Fig. 292.

Fig. 295.

Fig. 296.

Fig. 297.

Fig. 298.—Floating fenders at Dessouk Bridge, on the Nile.

Fig. 299.—Fenders at Bahr Shebin Bridge. Photograph taken before the span was erected.

Fig. 300.—Fenders of Mayiah Bridge with part of swing span above them.

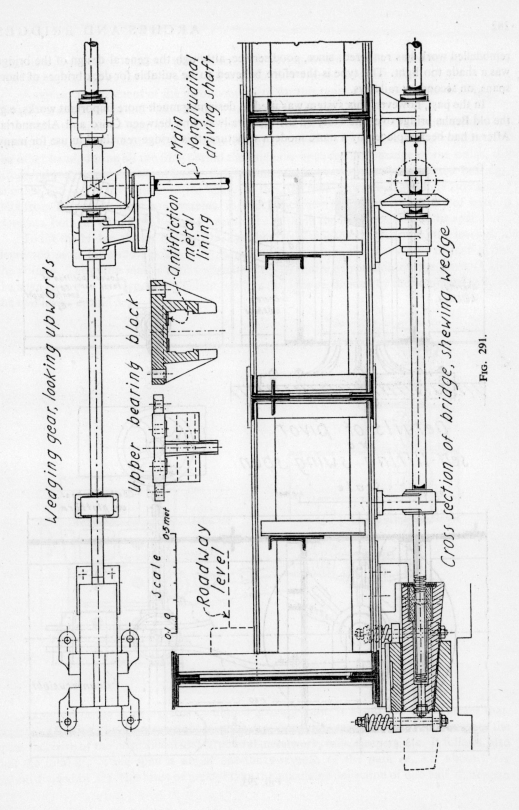

Fig. 291.

caused by its structural weight, plus the effect of the difference of temperatures of the upper and lower flanges.

The writer realizes that an objection against the presentation of the subject, as given in the foregoing pages, is that all examples are selected from Egyptian practice. In fact, it is dangerous to limit the horizon to types and methods as practised in one single country, for this may frequently lead to a one-sided approach.

It may, therefore, be of some interest to give an example of a wedging gear designed elsewhere than in Egypt, for instance, in the United States. Attention in this connection is called to Fig. 294, which shows a wedge in the Yorktown bridge.[1] It will be observed that whilst this is evidently a much larger work than illustrated in the foregoing pages, the principle of the design is generally the same.

4. Unusual Solutions

In perusing the foregoing pages, the reader may possibly come to the conclusion that irrigation bridges in Egypt, even the movable ones, are almost entirely standardized. Up to a point he may certainly be right, but this is not always the case, for the requirements of irrigation engineering may sometimes create altogether unusual bridge problems, calling, then, for unorthodox solutions and unusual bridge projects.

The bridge over the locks of the Mahmudieh Canal, in Alexandria, may be quoted as a case in point, in order to explain what is meant by the foregoing sentence. As seen from Fig. 295, it is built over two twin locks at the tail of the canal, where the latter discharges into the Alexandria harbour.

In addition to the locks, there is also an escape, for evacuating the surplus discharge.

Originally, there was a battery of four bridges: two road lift bridges of the type shown in Figs. 279 to 281, and two rail swing bridges as represented in Fig. 292. All these bridges were located within the precincts of the Alexandria Customs Administration and served as the chief and only link of communication between a large section of the customs area and its mainland. It was therefore felt, for some time past, that these light, standard structures were altogether inadequate to cope with the ever-increasing volume of traffic. But owing to the exiguity of the available space, and the necessity of keeping the road and rail traffic uninterrupted, and at the same time permit the boats to leave and enter the canal, no solution along traditional, classical lines was capable of being arrived at.

A rather unusual type was, therefore, adopted. All the four bridges were amalgamated into one single structure, which was a rim-bearing swing span, turning about a pivot located altogether eccentrically, on the left bank of the escape channel. The most conspicuous feature of the design was the patented rollers. As seen from Fig. 296, they were mounted in bogies, which were provided with secondary rollers, to allow for the thermal expansion of the span.

This was, in this case, a necessary precaution, because the alignment of the rolling path depended on the layout of the existing walls of the old locks, and the distance from the pivot to the rim was, consequently, much larger than in usual designs. Hence the necessity to provide for expansion at right angles to the rolling path.

In view of traffic requirements, erection by usual means was out of the question. About three-quarters of the steel-work was therefore erected in the open position of the span, as shown in Fig. 295, but the remaining part was mounted on two barges, and then floated into

[1] Abstracted from the paper "Special Design Features of the Yorktown Bridge", by Maurice N. Quade, published in *Trans. Am. Soc. C.E.*, Vol. 119, 1954, p. 119, Fig. 6.

AQUEDUCTS

First divide each span into three equal parts. This will give the points which are marked $a_1, a_2 \ldots$ in Diagram I.

Next, find the point b_1 at a distance $\dfrac{l_1}{3}$ from a_3. In a similar way, find b_2 and b_3, at distances $\dfrac{l_2}{3}$ and $\dfrac{l_3}{3}$ from the points a_3 and a_7 respectively.

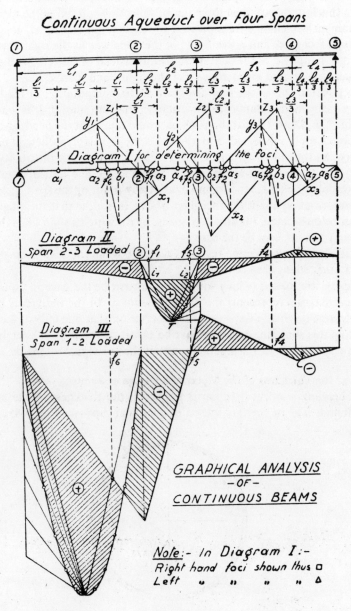

FIG. 303.

Draw an arbitrary line ①z_1, and then through the points y_1 and ② trace the line y_1 x_1. Join z_1 with x_1 and mark out the intersection, f_1, with the axis. This point possesses a particular value in the analysis of bending moments in continuous girders. It is therefore commonly referred to as "focus".

From the point f_1 draw another arbitrary line $f_1 z_2$ and find the point f_2 in the same way as was done in finding f_1. Then determine, similarly, point f_3. These points are called "right-hand" foci. If any load is placed on a span, to the right of the span considered, and there are no loads acting on this latter span, then the point of inflection will fall in the focus; that is to say, the bending moment will change from positive to negative, or *vice versa*, at such a point.

The left-hand foci f_4, f_5 and f_6 are found in the same way as the right-hand foci, but starting from the right-hand end of the continuous beam instead of the left-hand end. Their interpretation is analogous to that given above, for the right-hand foci.

When the positions of all the foci are known, the computation of the bending moment diagram is very easily made. Suppose, for instance, that the span ②③ is loaded as shown in Diagram II, Fig. 303. Draw the parabola which would represent the bending-moment diagram had the span been simply supported, and join the apex of this hypothetical parabola with the points ② and ③, by the lines T ② and T ③. Trace a line through the points i_1 and i_2 (intersections of T ② and T ③ with the verticals through f_1 and f_5). The vertical distance between this line and the parabola gives the bending moment at any point of the span ②③.

From the information about the foci given above, it is clear that the diagram of the bending moment can be completed, for all other spans, by drawing straight lines through the right-hand, or left-hand foci, as may be the case.

Diagram III in the same figure illustrates the case when the load is on one of the lateral spans, while all other spans are not loaded.

If all the spans are loaded (which will nearly always be the case in aqueducts), the final bending moment diagram represents the algebraic sum of all the individual bending-moment diagrams which were originally drawn on the assumption that one only of the spans was loaded. In calculating this sum, due account must be taken of the sign of each bending moment and the resultant calculated as the algebraic sum of the individual moments.

3. Calculation of the Thickness of the Metal and Design of Joints

When the bending-moment diagram is prepared, the thickness of the metal of the pipe may be easily found. The moment of inertia of a circular pipe (see Fig. 304) is:

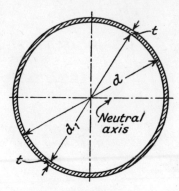

Fig. 304.

AQUEDUCTS

$$I = \frac{(d^4 - d_1^4)}{64} = 0.049\,(d^4 - d_1^4)$$

and the section modulus:

$$Z = \frac{(d^4 - d_1^4)}{32d} = \frac{0.098\,(d^4 - d_1^4)}{d}.$$

The application of these precise formulae in practice, involves calculations which, when the common slide-rule is used to perform the arithmetic, are beyond the numerical precision of the figures obtained; and therefore, the analytical precision of the solution does not improve the quantitative exactitude of the result.

Hence, the author finds that a simplified, but sufficiently approximate, method can be used instead; namely, we can neglect $\left(\frac{2t}{d_1}\right)^2$ as compared to $\frac{2t}{d}$ (see Fig. 304).

We will thus obtain:

$$Z = \frac{0.098\,d_1^4}{d}\left(\frac{d^4}{d_1^4} - 1\right) = \frac{0.098\,d_1^4}{d}\left[\left(1 + \frac{2t}{d_1}\right)^4 - 1\right].$$

Developing the binomial, we find:

$$Z = \text{approximately } 0.098\,d_1^3\,\frac{8t}{d_1} = \frac{\pi}{4}d^2 t,$$

wherefrom

$$t \geqslant \frac{4M}{f_s \pi d^2}$$

in which f_s is the permissible tensile stress limit.

For instance, let $D = 150$ cm and $t = 1$ cm. Then the exact formula gives:

$$Z = \frac{(152^4 - 150^4)}{32 \times 152} = \frac{(534{,}000{,}000 - 507{,}000{,}000)}{32 \times 152} = 17{,}430 \text{ cm}^3.$$

From the approximate formula we find:

$$Z = \frac{\pi}{4}\,150^2 \times 1 = 17{,}670 \text{ cm}^3.$$

The difference is $\frac{17{,}670 - 17{,}430}{17{,}430} = \frac{240}{17{,}430} = 1.4$ per cent,

which is quite sufficiently precise for practical purposes.

As a mnemonic rule, it is easy to remember that the *section modulus* of a pipe is equal to the *wetted area, multiplied by the thickness of the metal*. This transcribes, in words, the approximate formula which we have developed.[1]

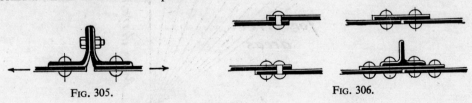

FIG. 305.　　　　　　　　　　　　　　　FIG. 306.

[1] As far as the author knows, this formula and mnemonic rule were first published by him in Volume II of his *Irrigation and Hydraulic Design*, page 801, in 1957.

As a general rule, the types of joints which are convenient for, and are frequently used in, syphons, should not be adopted in the design of aqueducts; because owing to longitudinal bending, the metal of a circular pipe in an aqueduct might be subjected to tension, which would cause this type of joint to open, as shown in Fig. 305.

The result would be: leaking of the joints and sagging of the pipe.

Four types of joints of various relative strengths are shown in Fig. 306. They possess different strengths but are all suitable for aqueducts. Alternatively, other types which can effectively resist tensile stresses can be employed.

The decision as to the number of rows of rivets in the joint, i.e. whether the joint will be a simple lap (or butt) or a double lap (or butt), will depend upon considerations regarding the shearing stress in the rivet shanks.

To find this stress, it is first necessary to calculate the moment of inertia of all the rivet sections, with reference to the horizontal neutral axis for bending.

This can be done by means of the graphical computation shown in Fig. 307. The shear area of every rivet, a, is assimilated to a stress, and a funicular polygon is then computed in the usual way, for this set of imaginary stresses. In this calculation it will suffice to consider one quarter of the pipe only.

The area, A, of the funicular polygon thus computed, multiplied by the polar distance,

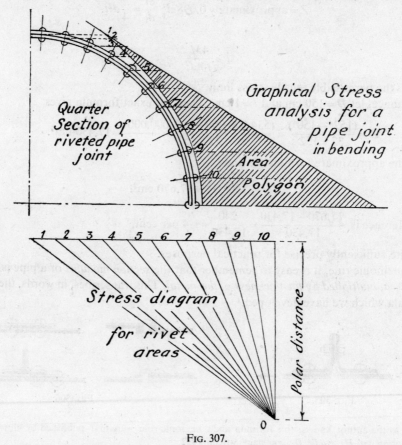

Fig. 307.

AQUEDUCTS

H, gives half the moment of inertia (second moment) of one-quarter of the rivets. It follows that

$$I = 8HA.$$

Note that the value of H is represented in the drawing to the same scale as is used in drawing the stress diagram; whereas A is to be measured by means of the planimeter, and the scale to be applied is that which was employed in drawing the section of the pipe.

Finally, in the case of a double row of rivets, we will use the formula $I = 16HA$.

If the position of the joint along the axis of the pipe is known, the bending moment may be easily found from the diagram of moments.

Then, the stress $f_s = \dfrac{Mr}{I}$ must be within the permissible limit for *shear*.

Here, again, as in the case of the plating, the calculation may be considerably simplified by means of an approximate, but nevertheless sufficiently precise, formula; in fact, imagine the row of rivets being replaced by an equivalent continuous circular band. Its width would then be equal to the sectional area, a, of the shank of the rivet divided by the pitch, p. Remembering our mnemonic rule, the section modulus of such a circular band would be approximately equal to $a\pi r^2/p$, and the shearing stress would therefore be

$$f_s = \frac{Mb}{a\pi r^2}.$$

In the case of a double row, this would change into

$$f_s = \frac{Mp}{2a\pi r^2}.$$

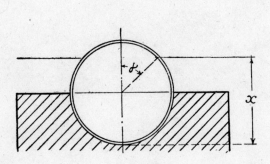

Fig. 308.

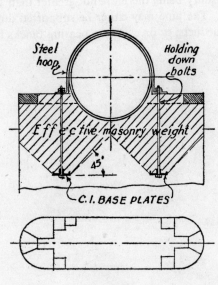

Fig. 309.

4. Anchorage

In order to discharge the water at the desired level, it may sometimes be necessary to place the pipe below the maximum water-level in the channel across which the aqueduct is built.

The upward load per unit length of aqueduct will then be equal to

$$\gamma \left[r^2 \pi \frac{180 + 2\alpha}{360} + (x-r)^2 \tan \alpha \right],$$

where x is the depth of the bottom of the pipe, below the maximum water-level (see Fig. 308), and γ the specific weight of water.

The anchorage will be designed on the assumption that the pipe is empty.

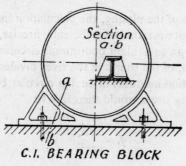

C.I. BEARING BLOCK
Fig. 310.

In calculating the true weight of masonry forming the ballast, as shown schematically in Fig. 309, we must take into account the loss due to buoyancy, the apparent specific weight of masonry being therefore not greater than 1.0.

The pipe may either be supported directly on masonry, as shown in Fig. 309, for small aqueducts; or on cast-iron bearing blocks for somewhat larger works, as shown in Fig. 310.

SUBJECT INDEX

Aqueducts
  bearing blocks for pipes, 294
  calculation of bending movement for pipe aqueducts, 288
  hydraulic design, 286
  number of rivets in joints, 292
  open *versus* pipe aqueducts, 286
  thickness of metal and structural design of joints, 291
  uplift (upward hydrostatic load) on pipe aqueducts, 294

Arches
  analysis of the three elastic equations, 8–15
  arbitrary assumptions for calculating an arch, 1–5
  arithmetical methods for solving elastic equations, 15–18
  Cochrane's method for arch design, 45–47
  deflection of curved beams, 5–9
  design procedure for an arch, 33–47
  elastic method (principle of) for the calculation of an arch, 4–5
  equations for thermal effect in arches, 30–31
  funicular polygon interpreted as the outline of the arch, 34–38
  graphical methods for solving elastic equations, 18–27
  inflection points for thermal movements, 31
  influence lines for arches, 18–23
  Kögler's method for arch design, 42–45
  Landsberg's solution, 25–26
  Langer's arch, 35
  Mehrtens' solution, 26–27
  Melan's method for arch design, 38–42
  reaction-curve method for arches, 23–27
  Schönhofer's method, 16–17
  stability of arches, in general, 1–5
  temperature stresses in arches, 27–33
  thermal movements in arches, 31
  three-hinged-arch principle, 4

Reinforced concrete canal bridges
  alignment charts for T-beams, 124–125
  analogy between a pre-stressed reinforced concrete beam and a retaining wall, 147
  bonded and non-bonded types of pre-stress, 145
  box-type *versus* H-type for pre-stressed beams, 159
  cross-girders in ordinary road bridge, 118
  details of standard steel reinforcement, 102–104
  evolution (historical) of pre-stress (according to Freyssinet), 142
  footpath (design of), 117
  form-factor (in pre-stress analysis), 151
  general principle of pre-stress, 140
  graphical method for the calculation of ordinary T-beams, 131–138
  influence lines for continuous beams, 118
  layouts for skew irrigation canal bridges, 157
  main girders in simply supported beam bridges, 128
  pre-stressed reinforced concrete, 140
  ribbed *versus* slab decks, 111
  slab (calculation of), 91, 112
  spacing of stirrups in beams, 128
  stresses in slab supported on four sides, 115
  table (abridged Gifford's) for design of pre-stressed beams, 163
  tables of constants for the calculation of ordinary slabs and beams, 95–98
  weight of steel in different types of canal bridges, 112
  wire specification (Magnel's) for pre-stressed bridges, 146

Rolled-joist canal bridges
  base plates and bearings for joists, 87
  calculation of reinforced concrete slab, 91
  compound type of beam, 107
  details of reinforced concrete slab, 102–104
  devices for fixing timber flooring, 84–90
  French formula for slab, 92
  pre-stressed (Dischinger's) type of compound beam, 110
  reinforced concrete flooring, 91
  steel girders embedded in concrete, 105
  tables of constants for calculation of depth of concrete and area of reinforcement, 95–98

tables of section moduli for rolled joists and channels (British, Continental and American), 85–89
timber flooring, 84
types with both steel and concrete participating in resisting bending moment, 106

Specifications for structural steel work (bridges and gates)
American impact allowances, 182
Association of American Railroads test, 180
bearing stress, as significant design criterion, 212
breaking load *versus* limit of stress criteria, 167
compressive stress limit, 207
difference in yield-points for tensile and bending stresses, 201
electro-magnetic strain recorder, 179
errors to be covered by safety factor, 192
formula for Launhardt-Weyrauch, 172
frequency period of normal oscillation, 176
graduated type of specifications for bridges, 174
impact type of specification *ditto*, 175
limit design principle and plastic theories, 198
limiting range for reversed stress, 171
model tests, carried out for the earliest types of metal bridges, 168
number of vehicles on span as function of their velocity, 185
photo-recording instrument (Fereday-Palmer), 177
ratio of shear to tensile limits, 211
rivets made in alloy steel, 189
rosette of stresses, 180
rust-resisting properties of various metalloids, 191
safety factor as a problem of mathematical probability, 195
shake-down principle, 204
SR-4 gauge, 179
statically-determined solution supplied by limit design principle, 199
stresses for alloy steel, 186
tensile limit as basic criterion, 166
Tepic gauge, 180
uncertainties *versus* ignorances in assessing safety factors, 195
wire-resistance type of strain recorder, 179

Steel canal bridges (calculation and design)
American type of wedging gear, 284
axle for a lift span, 265
bearings for fixed spans, 248–252
bending resistance of groups of rivets, 246
calculation of flange rivets, 221
continuous girders, 238–245
cross-girder design (load diagram), 231
devices (various) for fixing sleepers, 226–231
diagrams for the calculation of the section modulus and weight of a plate girder, 217–218
empirical formulas for the weight of bridges, 213
English *versus* Continental methods for plate-girder design, 215
fenders for swing bridges, 285
formulae for effective span when clear span is known, 213
general order of procedure in designing a steel bridge, 213
graphical methods for main girders, 237
lift bridges, 264–273
locking arrangement for swing spans, 276
rail bearers (stringers, joists, "*longerons sous voie*"), 225
road bridges of the plate-girder type, 256–260
rules of design for rivet connections, 260
Scherzer bridge, 267
sinusoid type of lift bridge, 270
spacing of sleepers, 229
Strauss bridge, 267
structural design of cross-girder, 232–234
symmetrical and asymmetrical designs of plate girders, 222–223
wedging gear for swing bridges, 275
welded connections, 261
wind bracings, 253

Timber canal bridges
angle between stress and fibre (effect of, on permissible stress limit), 49
built-up timber girders, 64
calculation and spacing of keys, 64–69
camber, 69
connectors, 73
design theory (classical assumptions), 51
flooring, 59
Golden Gate experiments, 74
homogeneity of timber, conditional on size of block, 78
Howe truss, 53
main girders in ordinary road bridge, 63
moisture effect on plywood beams, 78
new tendencies, deviating from classical theory, 73
permissible stress limits, 49
pre-stress in timber bridges, 71
shear plates, 73
slope of teeth, as affected by stress limits, 52
split rings, 73
standard live loads for road bridges, 59–61
stiffeners for plywood beams, 79
strengthening existing timber bridge, 71
trestles, 57
Tuchscherer system, 73
types of standard classical connections, 54

AUTHOR INDEX

Aston, J., 191
Aubry, C., 126

Bach, Prof., 102
Baker, Prof. A.L.L., 157
Baker, Prof. J.F., 198, 202
Baker, Sir, B., 174, 175
Barron, M., 157
Beard, 188
Broek, Prof. J.A. van den, 169, 198

Cochrane, V.H., 45 *et seq.*
Considère, 102
Coode, 273
Coulomb, 2
Culmann, 170

Dejardin, 40
Desnoyers, 40
Dietz, W., 267, 272
Dischinger, F., 110
Drucker, D.C., 169, 198, 205, 207

Ebeling, 80
Evans, R.H., 144

Faber, 142
Fairbairn, 168
Farid, T., 269
Feinberg, S., 169
Fereday, 177, 182
Freudenthal, Prof. A.M., 185, 194, 195, 197, 208
Freyssinet, E., 142, 143, 147, 152, 154, 165
Fuller, A.H., 185

Gerber, 170, 176
Gifford, F.W., 161, 163, 164
Glanville, 142

Hammill, H.B., 82
Hankinson, R.L., 50
Hansen, 77, 78, 79

Hedgkinson, 168
Hrennikoff, A., 169, 202
Huggenberger, 179, 180

Ilyushin, A.A., 198
Ivy, R.V., 175, 186

Jensen, Prof. V.P., 144
Johnson, T.H., 208

Karman, 209
Kelen, 33
Kempton-Dyson, H., 144
Ketchum, M.S., 93, 214
Kist, 198
Kögler, F., 42, 43, 45
Krohn, 209
Kuntze, W., 201

Landsberg, 173
Landsberg, Prof. T., 24, 25, 27
Langer, 35
Launhardt, 172, 173, 175
Lentze, 170
Levy, M., 198
Lewe, 73
Lin, T.V., 175

Magnel, Prof. G., 146
Mehrtens, Prof. G.C., 24, 27, 33
Melan, Prof., 74
Melan, Prof. Y., 37, 40, 41, 45
Mensch, L.J., 144
Mises, von, 198
Mitchell, 273
Mitcheil, S., 175
Mohr, 173
Moisseiff, 187

Nadai, A., 198
Navier, 50, 147, 148
Novak, 80

Olson, L.G., 82
Ostenfeld, 209

Palmer, 177, 182
Paul, A.A., 158, 159
Perronet, 40
Peterson, F.G.E., 201
Prager, W., 169, 198, 204
Prandtl, 198
Pugsley, Prof., 194

Quade, M.N., 284

Rankine, 40
Rebhan, 170
Richey, V.J., 175
Roab, N.C., 175
Robertson, Prof. R.G., 164
Roderik, J.W., 198
Roš, Prof. M., 35, 108
Ruble, E.J., 179, 180, 182

Saint-Venant, 198

Schaper, 249
Schaffey, C.F., 175
Schönhofer, 16 et seq., 47
Schneider, 176
Schwedler, 170
Séjourne, P., 24, 40
Stephenson, 168
Symonds, P., 204

Tetmajer, 208, 209
Thum, A., 210
Tuchscherer, 73

Vaughan-Lee, 273

Waddell, J.A.L., 185
Werder, 170
Weyrauch, Prof., 172, 173, 175, 193
Whetzel, J.C., 192
Wilson, 82, 273
Winkler, 27
Wöhler, 170, 171, 173
Wunderlich, F., 201